STRUCTURE, ROOFING
AND THE **EXTERIOR**

THE ILLUSTRATED HOME SERIES

ALSO AVAILABLE IN THE ILLUSTRATED HOME SERIES

HEATING AND AIR CONDITIONING

ELECTRICAL, PLUMBING, INSULATION AND THE INTERIOR

ALSO BY ALAN CARSON AND ROBERT DUNLOP

INSPECTING A HOUSE: A GUIDE FOR BUYERS, OWNERS, AND RENOVATORS

STRUCTURE, ROOFING AND THE EXTERIOR

THE ILLUSTRATED HOME SERIES

Carson Dunlop & Associates

Stoddart

Published in 2000 by Stoddart Publishing Co. Limited
34 Lesmill Road, Toronto, Canada M3B 2T6
180 Varick Street, 9th Floor, New York, New York 10014

Distributed in Canada by:
General Distribution Services Ltd.
325 Humber College Blvd., Toronto, Ontario M9W 7C3
Tel. (416) 213-1919   Fax (416) 213-1917
Email cservice@genpub.com

Distributed in the United States by:
General Distribution Services Inc.
PMB 128, 4500 Witmer Industrial Estates, Niagara Falls, New York 14305-1386
Toll-free Tel.1-800-805-1083   Toll-free Fax 1-800-481-6207
Email gdsinc@genpub.com

04   03   02   01   00   1   2   3   4   5

Canadian Cataloguing in Publication Data

Carson, Alan
Structure, roofing and the exterior
(Illustrated home series)
ISBN 0-7737-6145-4
1. Dwellings — Remodeling — Pictorial works. 2. Dwellings — Maintenance and repair — Pictorial works.
3. Roofs — Pictorial works. I. Dunlop, Robert. II. Title. III. Series: Carson, Alan. Illustrated home series
TH4816.C373 2000   690'.8'0286   C00-931476-8

U.S. Cataloguing in Publication Data
(Library of Congress Standards)

Carson, Alan.
Structure, roofing and the exterior / Alan Carson and Robert Dunlop. — 1st ed.
[128]p: ill. ; cm. (The Illustrated Home Series)
ISBN 0-7737-61454 (pbk.)
1. Siding (Building materials) — Amateurs' manuals. 2. Roofs — Maintenance and repair — Amateurs' manuals.
3. Dwellings —Maintenance and repair — Amateurs' manuals.
I. Dunlop, Robert. II. Title. III. Series
695 21 2000 CIP

Cover Design: Tannice Goddard
Text Illustrations: VECTROgraphics
Text Design: Neglia Design Inc./Tannice Goddard

THE CANADA COUNCIL | LE CONSEIL DES ARTS
FOR THE ARTS | DU CANADA
SINCE 1917 | DEPUIS 1957

*We acknowledge for their financial support of our
publishing program the Canada Council, the Ontario Arts
Council, and the Government of Canada through the
Book Publishing Industry Development Program (BPIDP).*

Printed and bound in Canada

CONTENTS

INTRODUCTION

At Carson Dunlop & Associates, a consulting engineering firm, our principal business is home inspection. A few years ago, we set out to build a distance education product for those planning to enter the home inspection business. We developed the Carson Dunlop Home Study System, the most comprehensive home inspection training program in existence.

Early in the development process, we were unable to find existing illustrations that would show our students what we were explaining to them in the text. Even when we could find good illustrations, they were limited to a specific topic, and similar illustrations for other subjects couldn't be found at all.

Enter Peter Yeates, an engineer at Carson Dunlop & Associates, and owner of a graphics company called VectroGraphics. He solved our problem by designing over 1,500 illustrations for us. Peter is one of a handful of people on the planet who combines the technical expertise, computer skills and aesthetics sense needed to create these illustrations.

Not only have these illustrations proven to be an effective tool in training home inspectors, but they have turned out to be very useful during home inspections. They allow us to explain situations to our clients, the potential homeowner.

The comments from students, other home inspectors and, most importantly, the home buying public, has led us to assemble the illustrations from the Home Study System and publish them as a series of books. This book, *Structure, Roofing, and the Exterior*, will give you all of the illustrations you'll need to understand those aspects of your house. We hope you will find this book useful, and that you'll also enjoy the other two books in the series, *Heating and Air Conditioning* and *Electrical, Plumbing, Insulation and the Interior*.

Finally, you don't have to feel guilty just looking at the pictures.

OUTSIDE THE HOUSE

This book covers the structural aspects of houses as well as exterior components, including the roof.

STRUCTURE

The structure is the most important part of a house. While the systems within the house are expected to age and fail, we expect the structure to last for hundreds of years.

The illustrations in this book will show you the various methods used for building a house from the ground up. In fact, some of the illustrations deal with the ground itself, as soil conditions can very often result in structural problems.

The structural illustrations follow a natural progression, starting with footings and foundations, moving on to floors and then wall systems, and finishing with the roof framing.

While the illustrations are self-explanatory, in some cases it may be beneficial to look at the series of illustrations that precede or follow the one you're concentrating on. They may show different methods of solving the same problem.

ROOFING

The roofing illustrations are divided into two main categories: steep roofs and flat roofs.

Steep roofs are shedding roofs. Shedding roofs are not watertight; they are more like a pyramid of umbrellas. The roofing material can vary from man-made products such as asphalt shingles to natural materials like wood and slate.

Most roofs leak where they change direction or run into another material. In these locations, flashings are used. The flashings are critical; you will find that many of the illustrations deal with flashing design and installation.

THE EXTERIOR

The exterior section shows architectural styles and features found in different homes.

Cladding is the technical name for the exterior "skin" of a building. The type of cladding depends on the age of the house and where it is located. There are several illustrations of cladding, as exterior walls can be covered by many materials, such as brick, masonry, stucco, wood, plywood, hardboard, and metal or vinyl siding. Some houses are even clad with asbestos cement siding, clay, slate, or insulbrick.

The exterior section also includes illustrations of doors and windows, as well as extensions of the house, such as porches, decks, balconies, carports, and garages.

While it is not part of the house proper, the landscaping and overall grading of the lot have a huge impact on basement dampness. The last section of the exterior deals with grading, gutters, downspouts, and window wells. The performance of all of these items dictates how wet (or dry) a basement will be.

INSTRUCTIONS FOR USE

You'll notice that there's a table of contents at the beginning of each part of this book. The heading of each illustration is listed next to the number of the illustration. Simply flip through the book until you find the illustration you want to look at.

PART 1

STRUCTURE

FOOTINGS AND FOUNDATIONS

DESCRIPTION

PROBLEMS

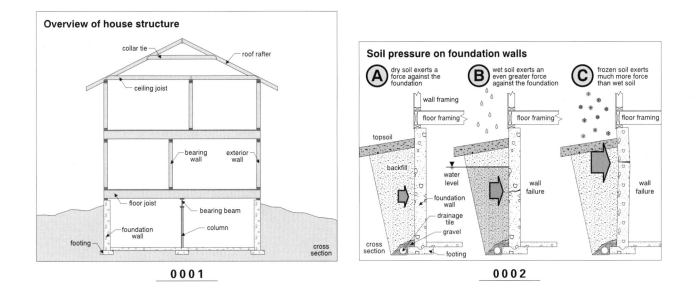

Overview of house structure

collar tie
roof rafter
ceiling joist
bearing wall
exterior wall
floor joist
bearing beam
foundation wall
column
footing
cross section

0001

Soil pressure on foundation walls

(A) dry soil exerts a force against the foundation
(B) wet soil exerts an even greater force against the foundation
(C) frozen soil exerts much more force than wet soil

wall framing
floor framing
floor framing
floor framing
topsoil
backfill
water level
foundation wall
drainage tile
gravel
wall failure
wall failure
cross section
footing

0002

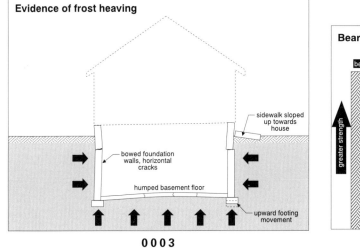

Evidence of frost heaving

sidewalk sloped up towards house
bowed foundation walls, horizontal cracks
humped basement floor
upward footing movement

0003

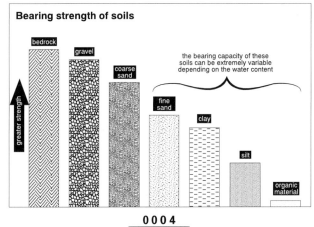

Bearing strength of soils

bedrock
gravel
coarse sand
the bearing capacity of these soils can be extremely variable depending on the water content
fine sand
clay
silt
organic material
greater strength

0004

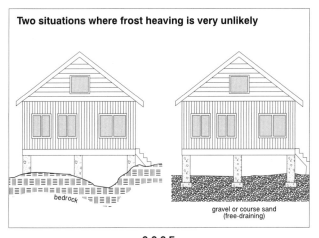

Two situations where frost heaving is very unlikely

bedrock
gravel or course sand (free-draining)

0005

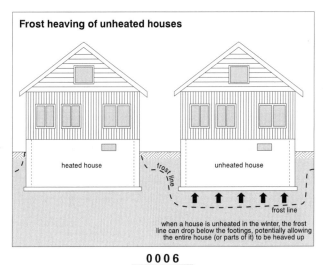

Frost heaving of unheated houses

heated house
frost line
unheated house
frost line

when a house is unheated in the winter, the frost line can drop below the footings, potentially allowing the entire house (or parts of it) to be heaved up

0006

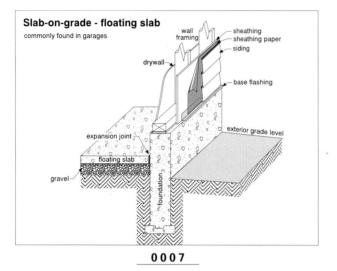

Slab-on-grade - floating slab
commonly found in garages

- wall framing
- sheathing
- sheathing paper
- siding
- drywall
- base flashing
- expansion joint
- exterior grade level
- floating slab
- foundation
- gravel

0007

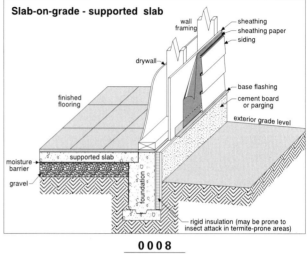

Slab-on-grade - supported slab

- wall framing
- sheathing
- sheathing paper
- siding
- drywall
- base flashing
- finished flooring
- cement board or parging
- exterior grade level
- supported slab
- moisture barrier
- gravel
- foundation
- rigid insulation (may be prone to insect attack in termite-prone areas)

0008

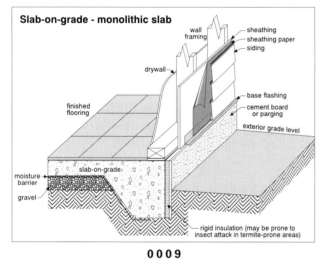

Slab-on-grade - monolithic slab

- wall framing
- sheathing
- sheathing paper
- siding
- drywall
- base flashing
- finished flooring
- cement board or parging
- exterior grade level
- slab-on-grade
- moisture barrier
- gravel
- rigid insulation (may be prone to insect attack in termite-prone areas)

0009

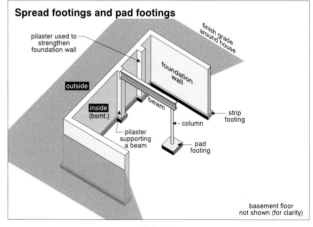

Spread footings and pad footings

- pilaster used to strengthen foundation wall
- finish grade around house
- foundation wall
- outside
- inside (bsmt.)
- beam
- strip footing
- column
- pilaster supporting a beam
- pad footing
- basement floor not shown (for clarity)

0010

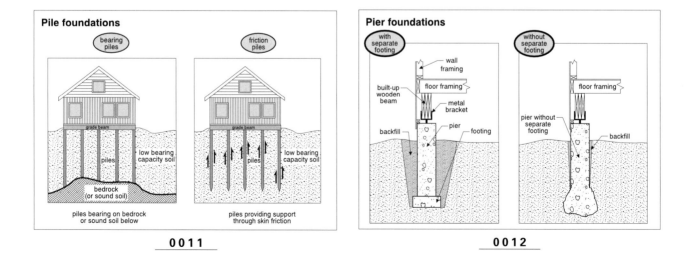

Pile foundations

- bearing piles
- friction piles
- grade beam
- piles
- low bearing capacity soil
- bedrock (or sound soil)
- piles bearing on bedrock or sound soil below
- grade beam
- piles
- low bearing capacity soil
- piles providing support through skin friction

0011

Pier foundations

- with separate footing
- without separate footing
- wall framing
- floor framing
- built-up wooden beam
- metal bracket
- backfill
- pier
- footing
- pier without separate footing
- floor framing
- backfill

0012

Brick foundation with masonry exterior walls

solid masonry exterior wall

floor framing

topsoil

backfill

brick foundation (typically 3 bricks thick)

brick footing

drainage tile may or may not be present

concrete floor slab

gravel

cross section

0013

Stone foundation with masonry exterior walls

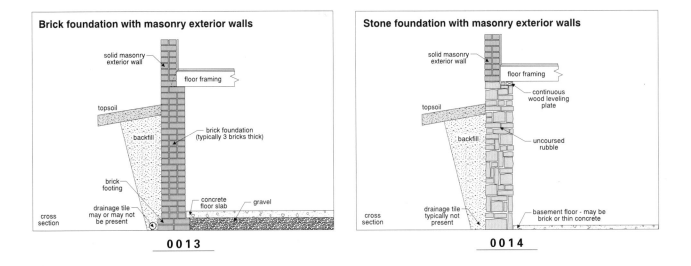

solid masonry exterior wall

floor framing

continuous wood leveling plate

topsoil

backfill

uncoursed rubble

drainage tile typically not present

basement floor - may be brick or thin concrete

cross section

0014

Stone foundation wall with wood frame exterior walls

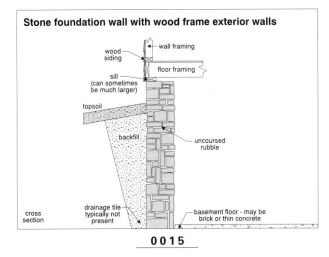

wall framing

wood siding

floor framing

sill (can sometimes be much larger)

topsoil

backfill

uncoursed rubble

drainage tile typically not present

basement floor - may be brick or thin concrete

cross section

0015

Raft and mat foundations

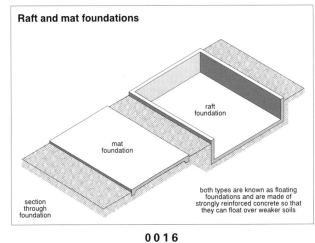

raft foundation

mat foundation

both types are known as floating foundations and are made of strongly reinforced concrete so that they can float over weaker soils

section through foundation

0016

Preserved wood foundations

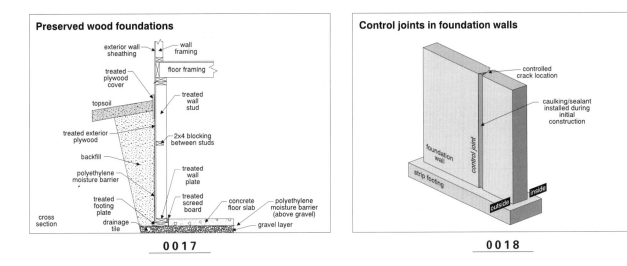

exterior wall sheathing

wall framing

treated plywood cover

floor framing

treated wall stud

topsoil

2x4 blocking between studs

treated exterior plywood

backfill

treated wall plate

polyethylene moisture barrier

treated footing plate

treated screed board

concrete floor slab

polyethylene moisture barrier (above gravel)

drainage tile

gravel layer

cross section

0017

Control joints in foundation walls

controlled crack location

caulking/sealant installed during initial construction

foundation wall

control joint

strip footing

outside

inside

outside

0018

Types of settlement

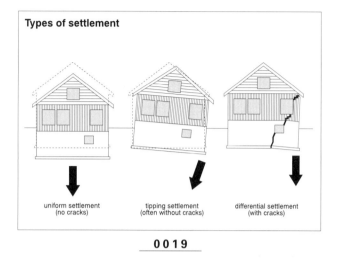

uniform settlement (no cracks)

tipping settlement (often without cracks)

differential settlement (with cracks)

0 0 1 9

Differential settlement caused by variable soil types

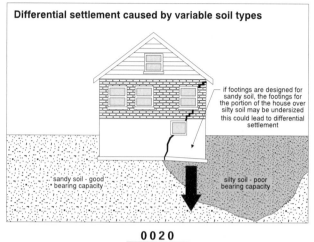

if footings are designed for sandy soil, the footings for the portion of the house over silty soil may be undersized this could lead to differential settlement

sandy soil - good bearing capacity

silty soil - poor bearing capacity

0 0 2 0

Ravine lots

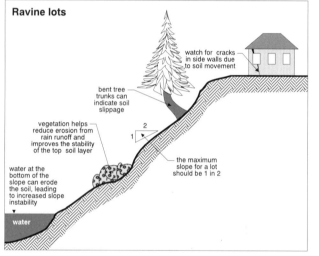

watch for cracks in side walls due to soil movement

bent tree trunks can indicate soil slippage

vegetation helps reduce erosion from rain runoff and improves the stability of the top soil layer

water at the bottom of the slope can erode the soil, leading to increased slope instability

water

2
1

the maximum slope for a lot should be 1 in 2

0 0 2 1

Building settlement due to cut and fill excavation

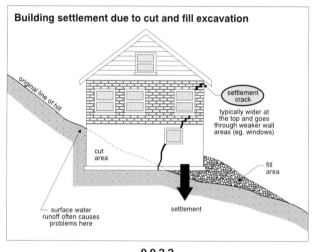

original line of hill

settlement crack

typically wider at the top and goes through weaker wall areas (eg. windows)

cut area

fill area

surface water runoff often causes problems here

settlement

0 0 2 2

Analyzing crack size

the size of individual cracks is not as important as the sum of all the crack sizes

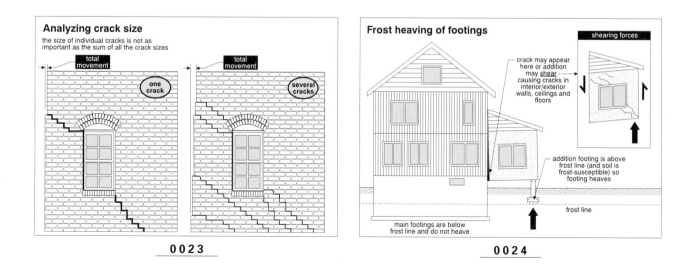

total movement

one crack

total movement

several cracks

0 0 2 3

Frost heaving of footings

shearing forces

crack may appear here or addition may shear - causing cracks in interior/exterior walls, ceilings and floors

addition footing is above frost line (and soil is frost-susceptible) so footing heaves

frost line

main footings are below frost line and do not heave

0 0 2 4

Crack shapes

V-cracks often indicate heaving and may be accompanied by crushing of material at the cracks

if two diagonal cracks form a pyramid, the pyramid is likely dropping

0025

Mud jacking to stabilize a settled foundation

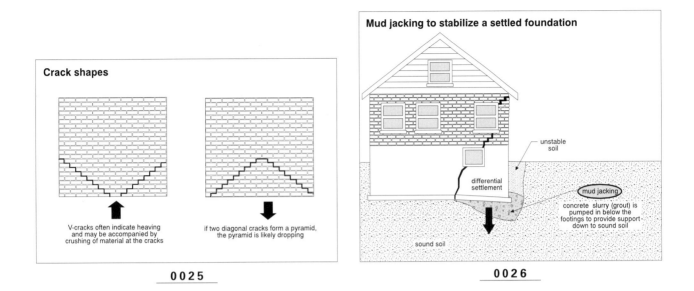

unstable soil

differential settlement

mud jacking

concrete slurry (grout) is pumped in below the footings to provide support down to sound soil

sound soil

0026

Using a helical anchor to stabilize a settled foundation

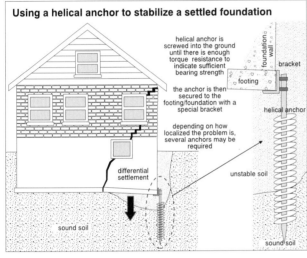

helical anchor is screwed into the ground until there is enough torque resistance to indicate sufficient bearing strength

foundation wall

bracket

footing

the anchor is then secured to the footing/foundation with a special bracket

helical anchor

depending on how localized the problem is, several anchors may be required

differential settlement

unstable soil

sound soil

sound soil

0027

Using piles to stabilize a settled foundation

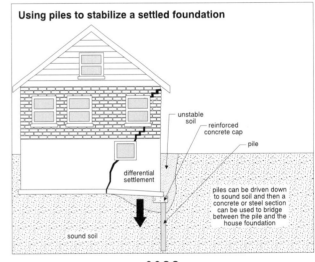

unstable soil

reinforced concrete cap

pile

differential settlement

piles can be driven down to sound soil and then a concrete or steel section can be used to bridge between the pile and the house foundation

sound soil

0028

Using rods and channels to stabilize a settled house

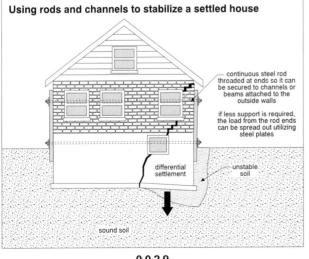

continuous steel rod threaded at ends so it can be secured to channels or beams attached to the outside walls

if less support is required, the load from the rod ends can be spread out utilizing steel plates

differential settlement

unstable soil

sound soil

0029

Step footings on sloped lots

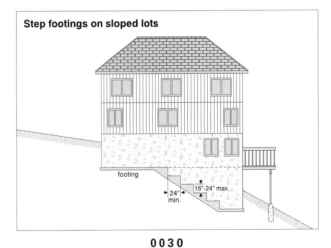

footing

24" min.

16"-24" max.

0030

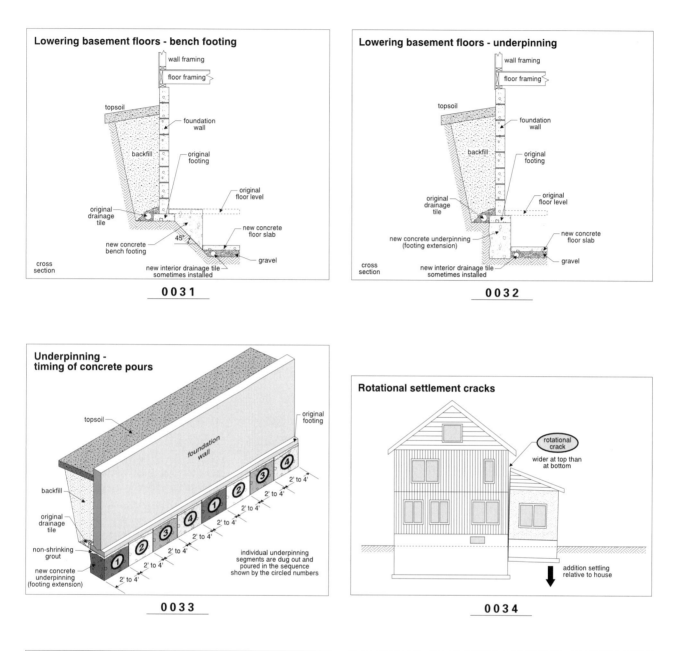

Lowering basement floors - bench footing

wall framing

floor framing

topsoil

foundation wall

backfill

original footing

original floor level

original drainage tile

new concrete floor slab

new concrete bench footing

45°

gravel

cross section

new interior drainage tile sometimes installed

0031

Lowering basement floors - underpinning

wall framing

floor framing

topsoil

foundation wall

backfill

original footing

original drainage tile

original floor level

new concrete underpinning (footing extension)

new concrete floor slab

gravel

cross section

new interior drainage tile sometimes installed

0032

Underpinning - timing of concrete pours

topsoil

original footing

foundation wall

backfill

original drainage tile

non-shrinking grout

new concrete underpinning (footing extension)

④ ③ ② ① ④ ③ ② ①

2' to 4'

individual underpinning segments are dug out and poured in the sequence shown by the circled numbers

0033

Rotational settlement cracks

rotational crack

wider at top than at bottom

addition settling relative to house

0034

Cracking - common locations

cracks often occur near high concentrated loads such as this settling column

cracks will also tend to show up at weak areas such as the line through the doors and windows

0035

Frost heaving of footings

crack may appear here or addition may shear - causing cracks in interior/exterior walls, ceilings and floors

shearing forces

addition footing is above frost line (and soil is frost-susceptible) so footing heaves

frost line

main footings are below frost line and do not heave

0036

Sources of heaving

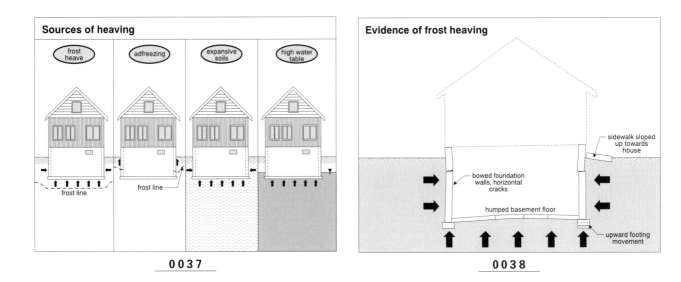

frost heave | adfreezing | expansive soils | high water table

frost line

frost line

0037

Evidence of frost heaving

sidewalk sloped up towards house

bowed foundation walls, horizontal cracks

humped basement floor

upward footing movement

0038

Exterior basement stairwell

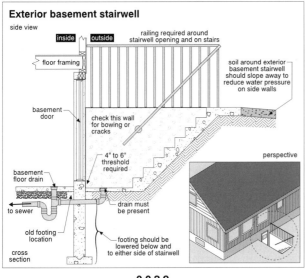

side view

inside | outside

railing required around stairwell opening and on stairs

floor framing

soil around exterior basement stairwell should slope away to reduce water pressure on side walls

basement door

check this wall for bowing or cracks

4" to 6" threshold required

perspective

basement floor drain

to sewer

old footing location

drain must be present

footing should be lowered below and to either side of stairwell

cross section

0039

Insulated exterior basement stairwell

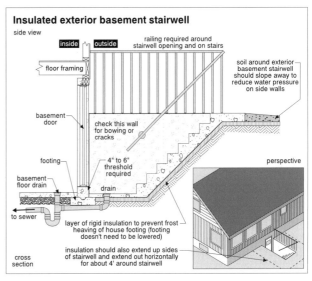

side view

inside | outside

railing required around stairwell opening and on stairs

floor framing

soil around exterior basement stairwell should slope away to reduce water pressure on side walls

basement door

check this wall for bowing or cracks

footing

4" to 6" threshold required

perspective

basement floor drain

drain

to sewer

layer of rigid insulation to prevent frost heaving of house footing (footing doesn't need to be lowered)

insulation should also extend up sides of stairwell and extend out horizontally for about 4' around stairwell

cross section

0040

Frost heaving of unheated houses

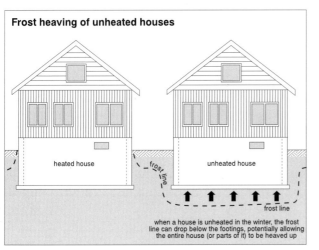

heated house

unheated house

frost line

frost line

when a house is unheated in the winter, the frost line can drop below the footings, potentially allowing the entire house (or parts of it) to be heaved up

0041

Effects of adfreezing

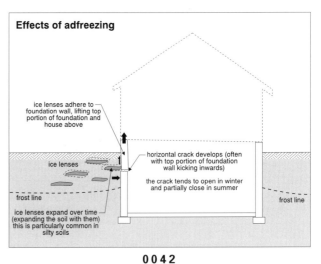

ice lenses adhere to foundation wall, lifting top portion of foundation and house above

horizontal crack develops (often with top portion of foundation wall kicking inwards)

the crack tends to open in winter and partially close in summer

ice lenses

frost line

frost line

ice lenses expand over time (expanding the soil with them) this is particularly common in silty soils

0042

Adfreezing and deck piers

even if the frost depth is above the bottom of the deck pier, soil adfreezing to the upper portion can cause heaving

sometimes plastic slip sheets are installed around the pier to prevent the frozen soil from gripping the concrete

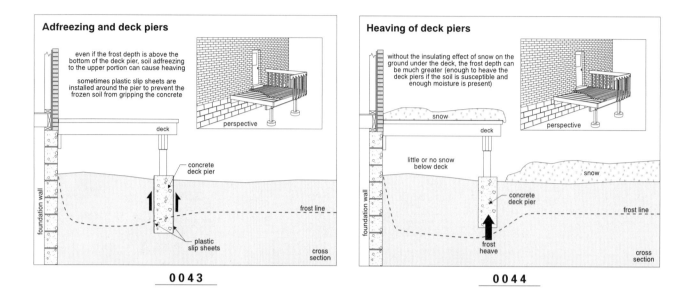

perspective

deck

concrete deck pier

foundation wall

frost line

plastic slip sheets

cross section

0043

Heaving of deck piers

without the insulating effect of snow on the ground under the deck, the frost depth can be much greater (enough to heave the deck piers if the soil is susceptible and enough moisture is present)

perspective

snow

deck

little or no snow below deck

snow

foundation wall

concrete deck pier

frost line

frost heave

cross section

0044

Foundation movement associated with horizontal cracks

horizontal foundation cracks are often accompanied by bowing, bulging or leaning

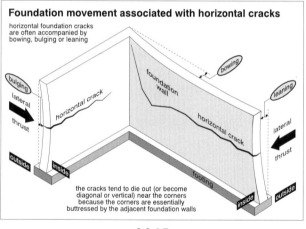

bowing

bulging

lateral thrust

foundation wall

horizontal crack

horizontal crack

leaning

lateral thrust

outside

inside

footing

inside

outside

the cracks tend to die out (or become diagonal or vertical) near the corners because the corners are essentially buttressed by the adjacent foundation walls

0045

Floors provide lateral support for foundations

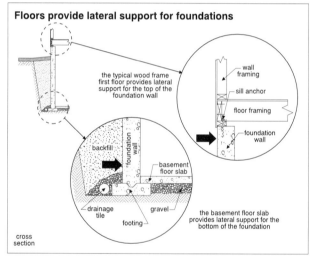

the typical wood frame first floor provides lateral support for the top of the foundation wall

wall framing

sill anchor

floor framing

foundation wall

backfill

foundation wall

basement floor slab

drainage tile

gravel

footing

the basement floor slab provides lateral support for the bottom of the foundation

cross section

0046

Horizontal cracks and movement - 3 different possibilities

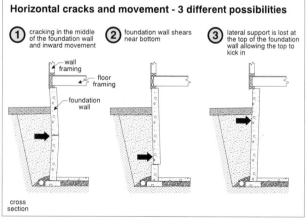

① cracking in the middle of the foundation wall and inward movement

② foundation wall shears near bottom

③ lateral support is lost at the top of the foundation wall allowing the top to kick in

wall framing

floor framing

foundation wall

cross section

0047

Beam punching through foundation wall

with sufficient lateral force, the foundation wall can bow inward so much that the end of the beam punches through the top of the foundation

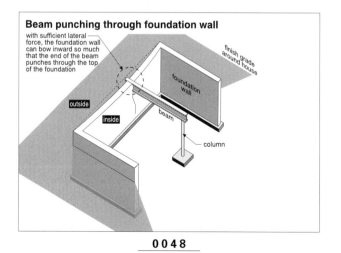

finish grade around house

foundation wall

outside

inside

beam

column

0048

Excess brick overhang resulting from bowing foundation

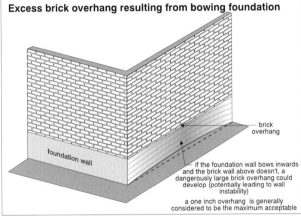

brick overhang

foundation wall

if the foundation wall bows inwards and the brick wall above doesn't, a dangerously large brick overhang could develop (potentially leading to wall instability)

a one inch overhang is generally considered to be the maximum acceptable

0049

Lateral support for foundation walls

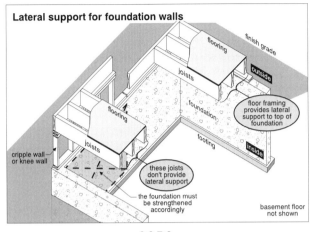

finish grade
flooring
joists
outside
foundation
floor framing provides lateral support to top of foundation
footing
inside
flooring
joists
cripple wall or knee wall
these joists don't provide lateral support
the foundation must be strengthened accordingly
basement floor not shown

0050

Vertical foundation cracks

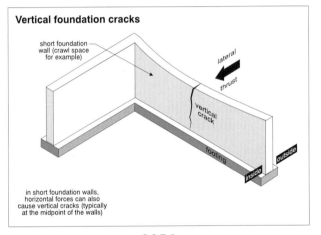

short foundation wall (crawl space for example)

lateral thrust

vertical crack

footing

inside　outside

in short foundation walls, horizontal forces can also cause vertical cracks (typically at the midpoint of the walls)

0051

Determining height of backfill

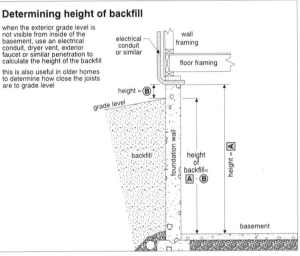

when the exterior grade level is not visible from inside of the basement, use an electrical conduit, dryer vent, exterior faucet or similar penetration to calculate the height of the backfill

this is also useful in older homes to determine how close the joists are to grade level

wall framing
electrical conduit or similar
floor framing
height = (B)
grade level
backfill
height of backfill= A - B
foundation wall
height = A
basement

0052

How driveways can contribute to foundation cracking

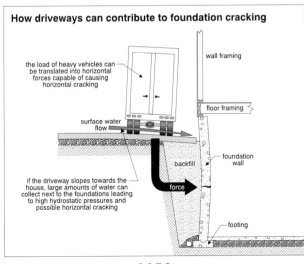

the load of heavy vehicles can be translated into horizontal forces capable of causing horizontal cracking

wall framing
floor framing
surface water flow
backfill
foundation wall
force
footing

if the driveway slopes towards the house, large amounts of water can collect next to the foundations leading to high hydrostatic pressures and possible horizontal cracking

0053

Foundation cracks related to tree roots

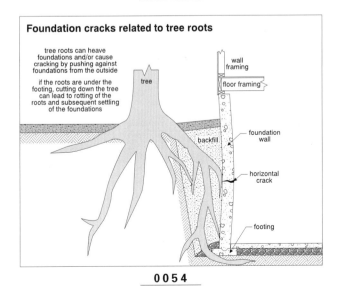

tree roots can heave foundations and/or cause cracking by pushing against foundations from the outside

if the roots are under the footing, cutting down the tree can lead to rotting of the roots and subsequent settling of the foundations

tree
wall framing
floor framing
backfill
foundation wall
horizontal crack
footing

0054

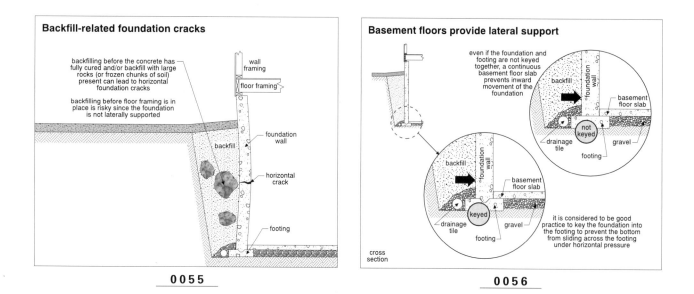

Backfill-related foundation cracks

backfilling before the concrete has fully cured and/or backfill with large rocks (or frozen chunks of soil) present can lead to horizontal foundation cracks

backfilling before floor framing is in place is risky since the foundation is not laterally supported

wall framing

floor framing

backfill

foundation wall

horizontal crack

footing

0055

Basement floors provide lateral support

even if the foundation and footing are not keyed together, a continuous basement floor slab prevents inward movement of the foundation

backfill

foundation wall

basement floor slab

not keyed

drainage tile

gravel

footing

backfill

foundation wall

basement floor slab

keyed

drainage tile

gravel

footing

it is considered to be good practice to key the foundation into the footing to prevent the bottom from sliding across the footing under horizontal pressure

cross section

0056

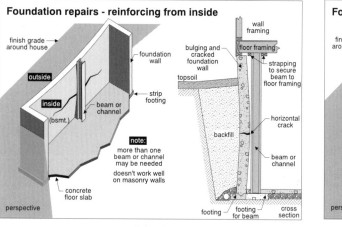

Foundation repairs - adding a buttress

finish grade around house

foundation wall

outside

inside

(bsmt.)

buttress

securing the buttress to the floor framing increases its strength

concrete floor slab

perspective

wall framing

floor framing

bulging and cracked foundation wall

topsoil

buttress

horizontal crack

backfill

additional buttress footing (if req'd)

footing

cross section

strip footing

0057

Foundation repairs - adding a pilaster

finish grade around house

foundation wall

outside

inside

(bsmt.)

pilaster

cores of pilaster blocks would ideally be filled with concrete and reinforced

securing the pilaster to the floor framing would also increase its strength

concrete floor slab

perspective

wall framing

floor framing

bulging and cracked foundation wall

topsoil

pilaster

horizontal crack

backfill

additional pilaster footing (if req'd)

footing

cross section

strip footing

0058

Foundation repairs - reinforcing from inside

finish grade around house

foundation wall

outside

inside

(bsmt.)

beam or channel

note:
more than one beam or channel may be needed

doesn't work well on masonry walls

concrete floor slab

perspective

wall framing

floor framing

bulging and cracked foundation wall

topsoil

strapping to secure beam to floor framing

horizontal crack

backfill

beam or channel

footing

footing for beam

cross section

strip footing

0059

Foundation repairs - using steel tie-backs

finish grade around house

foundation wall

outside

inside

horizontal crack

(bsmt.)

note:
not effective on masonry walls

concrete floor slab

perspective

wall framing

floor framing

bulging and cracked foundation wall

topsoil

anchor plates and nuts

horizontal crack

backfill

steel tie rods

hinged anchors

footing

cross section

strip footing

0060

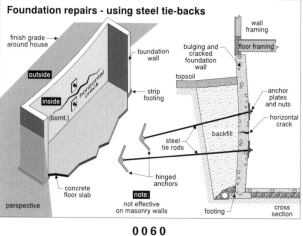

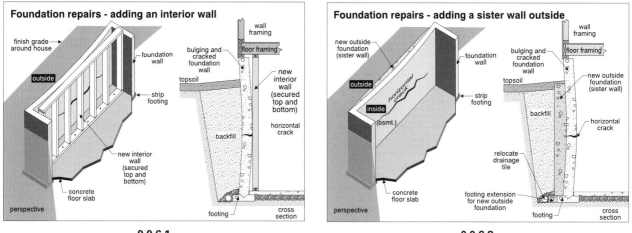

Foundation repairs - adding an interior wall

finish grade around house

foundation wall

outside

strip footing

new interior wall (secured top and bottom)

concrete floor slab

perspective

wall framing

floor framing

bulging and cracked foundation wall

topsoil

new interior wall (secured top and bottom)

backfill

horizontal crack

footing

cross section

0061

Foundation repairs - adding a sister wall outside

new outside foundation (sister wall)

foundation wall

outside

inside (bsmt.)

horizontal crack

strip footing

concrete floor slab

perspective

wall framing

floor framing

bulging and cracked foundation wall

topsoil

new outside foundation (sister wall)

backfill

horizontal crack

relocate drainage tile

footing extension for new outside foundation

footing

cross section

0062

Spalling of poured concrete foundations

wall framing

floor framing

topsoil

backfill

spalling or deterioration of the top of the foundation wall can cause serious loss of support for floors and walls

poor quality concrete is prone to spalling

if the amount of material lost is significant, a specialist should be consulted

footing

concrete floor slab

cross section

0063

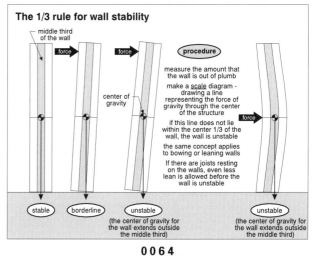

The 1/3 rule for wall stability

middle third of the wall

force

force

center of gravity

procedure

measure the amount that the wall is out of plumb

make a scale diagram - drawing a line representing the force of gravity through the center of the structure

force

if this line does not lie within the center 1/3 of the wall, the wall is unstable

the same concept applies to bowing or leaning walls

If there are joists resting on the walls, even less lean is allowed before the wall is unstable

stable

borderline

unstable (the center of gravity for the wall extends outside the middle third)

unstable (the center of gravity for the wall extends outside the middle third)

0064

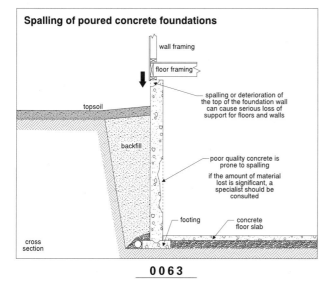

Piers - things to watch for

if the beam or joist has insufficient bearing, crushing of the wood and deflection can occur

beam or joist

the eccentric loading can also cause the pier to rotate out of plumb

pier

joist

beam

pier

the top of the pier should be wide enough to support the full width of the beam or shearing forces can cause deflection

0065

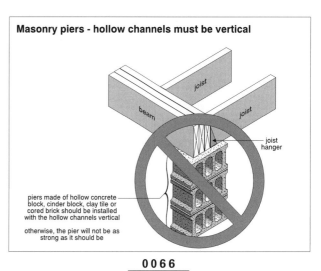

Masonry piers - hollow channels must be vertical

joist

beam

joist

joist hanger

piers made of hollow concrete block, cinder block, clay tile or cored brick should be installed with the hollow channels vertical

otherwise, the pier will not be as strong as it should be

0066

Wooden piers - things to look for

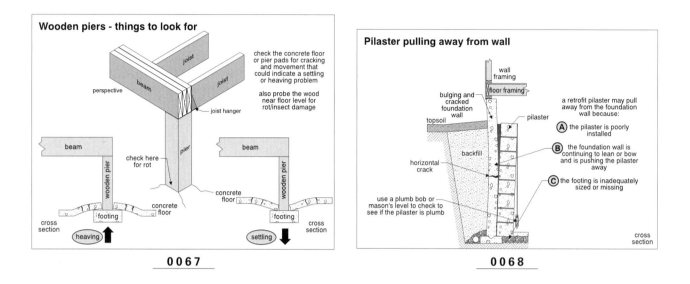

perspective

joist

beam

joist

joist hanger

check the concrete floor or pier pads for cracking and movement that could indicate a settling or heaving problem

also probe the wood near floor level for rot/insect damage

beam

wooden pier

check here for rot

pier

beam

wooden pier

concrete floor

concrete floor

footing

footing

cross section

cross section

heaving

settling

0 0 6 7

Pilaster pulling away from wall

wall framing

floor framing

bulging and cracked foundation wall

topsoil

pilaster

a retrofit pilaster may pull away from the foundation wall because:

(A) the pilaster is poorly installed

(B) the foundation wall is continuing to lean or bow and is pushing the pilaster away

(C) the footing is inadequately sized or missing

horizontal crack

backfill

use a plumb bob or mason's level to check to see if the pilaster is plumb

cross section

0 0 6 8

Cold joints in poured concrete foundations

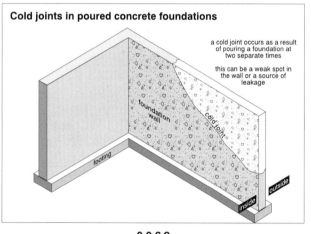

a cold joint occurs as a result of pouring a foundation at two separate times

this can be a weak spot in the wall or a source of leakage

foundation wall

cold joint

footing

outside

inside

0 0 6 9

Honeycombing in poured concrete foundations

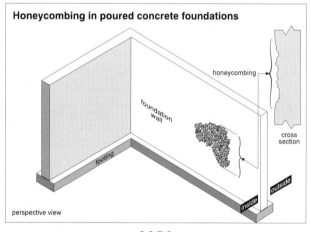

honeycombing

foundation wall

cross section

footing

perspective view

inside

outside

0 0 7 0

Foundation wall too short (soil level too high)

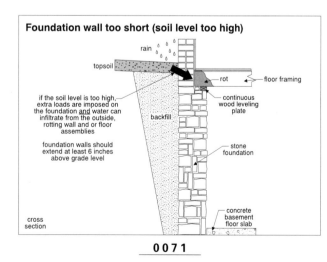

rain

topsoil

rot

floor framing

if the soil level is too high, extra loads are imposed on the foundation and water can infiltrate from the outside, rotting wall and or floor assemblies

continuous wood leveling plate

backfill

foundation walls should extend at least 6 inches above grade level

stone foundation

cross section

concrete basement floor slab

0 0 7 1

Lateral support for foundation walls

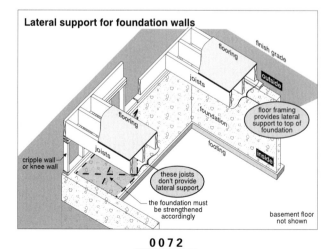

flooring

finish grade

joists

outside

flooring

foundation

floor framing provides lateral support to top of foundation

joists

footing

cripple wall or knee wall

inside

these joists don't provide lateral support

the foundation must be strengthened accordingly

basement floor not shown

0 0 7 2

Allowable floor deflections

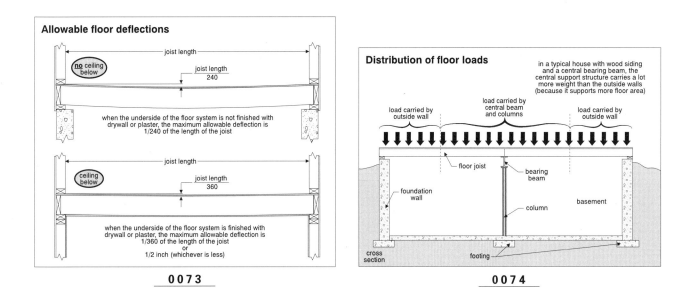

joist length

no ceiling below

joist length
240

when the underside of the floor system is not finished with drywall or plaster, the maximum allowable deflection is 1/240 of the length of the joist

joist length

ceiling below

joist length
360

when the underside of the floor system is finished with drywall or plaster, the maximum allowable deflection is 1/360 of the length of the joist
or
1/2 inch (whichever is less)

0073

Distribution of floor loads

in a typical house with wood siding and a central bearing beam, the central support structure carries a lot more weight than the outside walls (because it supports more floor area)

load carried by outside wall

load carried by central beam and columns

load carried by outside wall

floor joist

bearing beam

foundation wall

column

basement

cross section

footing

0074

Air gaps around ends of wooden beams

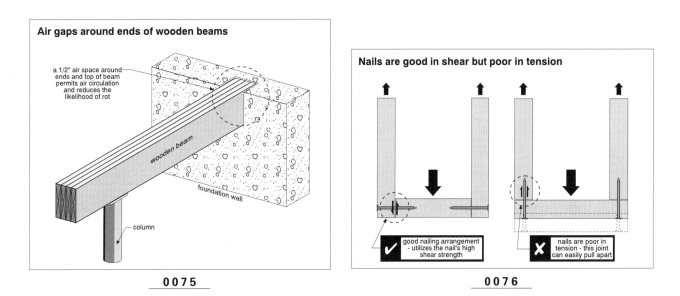

a 1/2" air space around ends and top of beam permits air circulation and reduces the likelihood of rot

wooden beam

foundation wall

column

0075

Nails are good in shear but poor in tension

✓ good nailing arrangement - utilizes the nail's high shear strength

✗ nails are poor in tension - this joint can easily pull apart

0076

Sills should be above grade

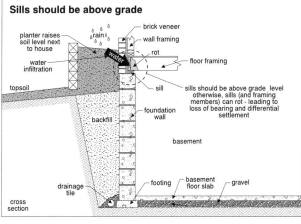

planter raises soil level next to house

rain

brick veneer

wall framing

water

rot

floor framing

water infiltration

topsoil

sill

sills should be above grade level otherwise, sills (and framing members) can rot - leading to loss of bearing and differential settlement

backfill

foundation wall

basement

drainage tile

footing

basement floor slab

gravel

cross section

0077

Mud sills

wall sheathing

siding

wall framing

floor sheathing

floor framing

crawlspace

mud sill

mud sill

outside grade

the wooden mud sill found in some older homes is prone to rot and insect attack and should be carefully inspected

0078

Sill anchor spacing

anchor bolts should be on 8 foot centers (Canada) <u>or</u> on 6 foot centers (USA)

bolts are needed within 12 inches of corners (USA)

wall framing

floor framing

foundation wall

1/2" anchor bolt

nut

washer

sill gasket

sill

4" to 6"

foundation wall

0079

Gaps under sills

if the top of the foundation is not even, the sill (and the floor and wall systems) can deflect once they are loaded

a bed of mortar is sometimes used to level the tops of uneven foundations

floor sheathing

floor framing

floor joists

sill

foundation wall

studs

foundation wall

view facing foundation

perspective view

foundation wall

note:
unevenness at top of foundation exaggerated for clarity

0080

Sill crushing

floor joist

floor joist

sill

foundation wall

look for crushing of the sills at the ends of the joists

this is more likely to be a problem where the sills are near grade level (rot problems) and/or where the joists have too little end bearing (concentrated load)

perspective view
(joist shown semi-transparent for better visibility)

0081

Column types

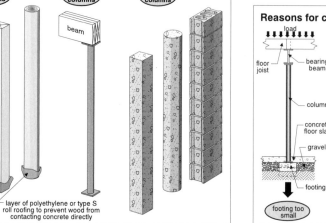

wood columns

steel columns

concrete columns

beam

beam

layer of polyethylene or type S roll roofing to prevent wood from contacting concrete directly

0082

Reasons for column settling

load

floor joist

bearing beam

column

concrete floor slab

gravel

footing

footing too small

no footing

poor soil conditions under footing

larger than intended load on column

0083

Column crushing

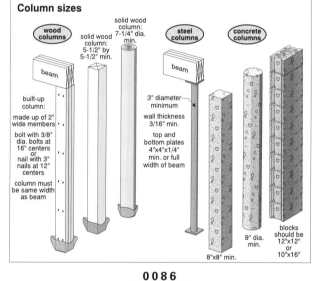

wood shims are often used at the top of columns

these may be cedar wedges (weaker wood and small bearing surfaces) that are particularly prone to crushing

wedges between column and beam mean that the beam is not well connected to the column

floor joist

built-up beam

concrete block column

footing

concrete floor slab

gravel

floor joist

built-up beam

wood column

footing

concrete floor slab

gravel

with wooden columns, check for crushing at both the top <u>and</u> the bottom

if there is crushing at the bottom, rot may be involved

0084

The 1/3 rule for column stability

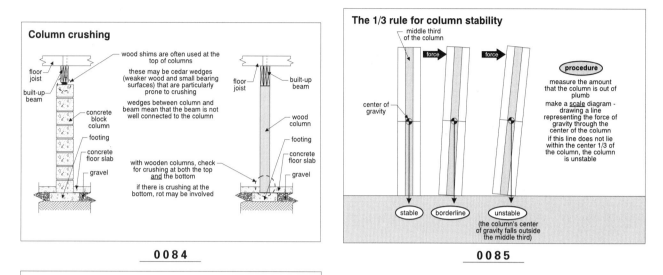

middle third of the column

force

force

center of gravity

procedure

measure the amount that the column is out of plumb

make a <u>scale</u> diagram - drawing a line representing the force of gravity through the center of the column

if this line does not lie within the center 1/3 of the column, the column is unstable

stable

borderline

unstable

(the column's center of gravity falls outside the middle third)

0085

Column sizes

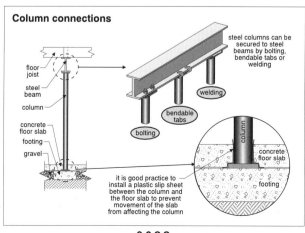

wood columns

solid wood column: 5-1/2" by 5-1/2" min.

solid wood column: 7-1/4" dia. min.

steel columns

concrete columns

beam

beam

built-up column:

made up of 2" wide members

bolt with 3/8" dia. bolts at 16" centers
or
nail with 3" nails at 12" centers

column must be same width as beam

3" diameter minimum

wall thickness 3/16" min.

top and bottom plates 4"x4"x1/4" min. or full width of beam

8"x8" min.

9" dia. min.

blocks should be 12"x12" or 10"x16"

0086

Column connections

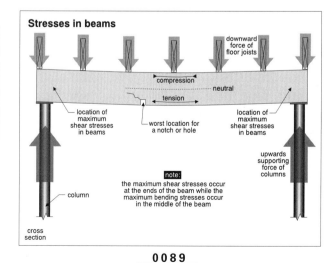

floor joist

steel beam

column

concrete floor slab

footing

gravel

steel columns can be secured to steel beams by bolting, bendable tabs or welding

welding

bendable tabs

bolting

column

concrete floor slab

footing

it is good practice to install a plastic slip sheet between the column and the floor slab to prevent movement of the slab from affecting the column

0088

Column buckling

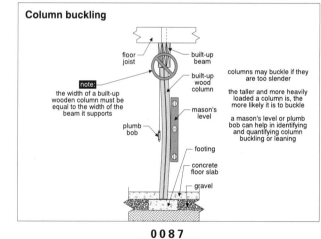

floor joist

built-up beam

built-up wood column

note:
the width of a built-up wooden column must be equal to the width of the beam it supports

mason's level

plumb bob

footing

concrete floor slab

gravel

columns may buckle if they are too slender

the taller and more heavily loaded a column is, the more likely it is to buckle

a mason's level or plumb bob can help in identifying and quantifying column buckling or leaning

0087

Stresses in beams

downward force of floor joists

compression

neutral

tension

location of maximum shear stresses in beams

worst location for a notch or hole

location of maximum shear stresses in beams

upwards supporting force of columns

column

cross section

note:
the maximum shear stresses occur at the ends of the beam while the maximum bending stresses occur in the middle of the beam

0089

Load transfer

joists supported on top of beam

joists supported on bottom flange of steel beam

beam

joist

beam

joist

perspective view

0090

Mounting joists onto steel beams

inferior   sometimes found in older homes - (causes joist-weakening cracks)

beam

joist

crack extends from notch in joist

joist

front view

perspective view

preferred

2"x2" splice at least 2' long

provide 1/2" clearance between splice and beam to accommodate shrinkage

joist   joist

beam

joist   joist

beam

continuous 2"x2" bolted through beam web every 2' (min.) with 1/4" dia. bolts

front view

0091

Continuous beams are stiffer than simply spanned beams

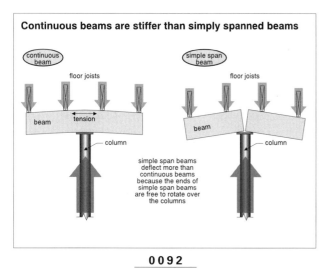

continuous beam

floor joists

simple span beam

floor joists

beam   tension

column

beam

column

simple span beams deflect more than continuous beams because the ends of simple span beams are free to rotate over the columns

0092

Sight along beams to check for sag

even with a mason's level, it can be difficult to determine if a beam is sagging by looking at it from the side

sighting along the bottom edge of the beam gives a much clearer indication of whether it's sagging

side view

perspective view

0093

Support for beam ends

finish grade around house

foundation wall

outside

inside

beam

column

pilaster supporting beam

foundation wall

steel beam

3-1/2" min. bearing

steel beams and columns should not bear on wood (or pieces of brick)

steel plate shims are most appropriate for levelling the end of a beam (they should be welded together and to the beam)

perspective

basement floor not shown (for clarity)

0094

Rotated or twisted beams

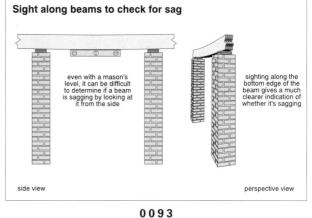

floor joist

built-up beam

masonry column

beam rotation can cause point bearing situations leading to localized crushing

0095

Notches or holes not allowed in beams

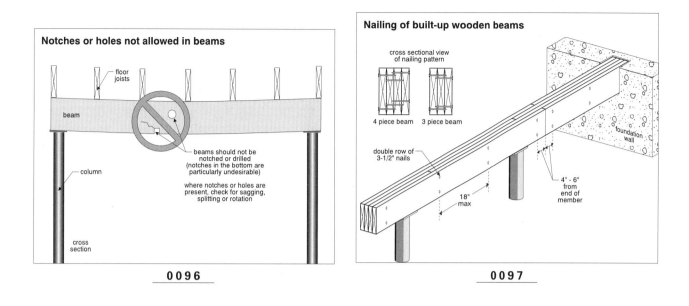

floor joists

beam

beams should not be notched or drilled (notches in the bottom are particularly undesirable)

where notches or holes are present, check for sagging, splitting or rotation

column

cross section

0096

Nailing of built-up wooden beams

cross sectional view of nailing pattern

4 piece beam   3 piece beam

double row of 3-1/2" nails

foundation wall

18" max

4" - 6" from end of member

0097

Bolting of built-up wooden beams

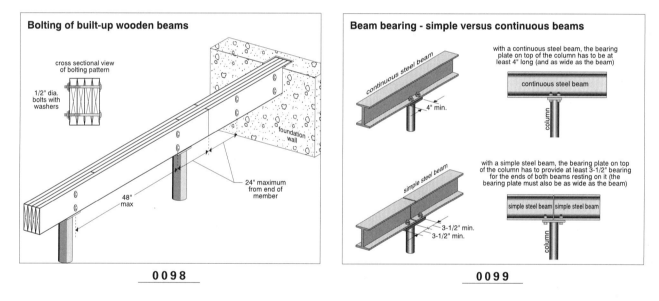

cross sectional view of bolting pattern

1/2" dia. bolts with washers

foundation wall

48" max

24" maximum from end of member

0098

Beam bearing - simple versus continuous beams

with a continuous steel beam, the bearing plate on top of the column has to be at least 4" long (and as wide as the beam)

continuous steel beam

4" min.

continuous steel beam

column

with a simple steel beam, the bearing plate on top of the column has to provide at least 3-1/2" bearing for the ends of both beams resting on it (the bearing plate must also be as wide as the beam)

simple steel beam

3-1/2" min.
3-1/2" min.

simple steel beam   simple steel beam

column

0099

Column connections

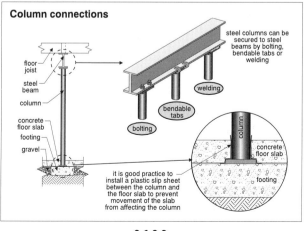

steel columns can be secured to steel beams by bolting, bendable tabs or welding

floor joist

steel beam

column

concrete floor slab

footing

gravel

welding

bendable tabs

bolting

column

concrete floor slab

footing

it is good practice to install a plastic slip sheet between the column and the floor slab to prevent movement of the slab from affecting the column

0100

Examples of weak joist/beam connections

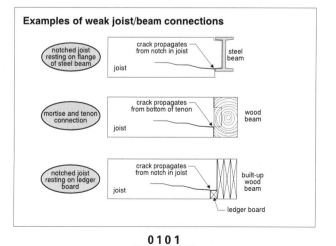

notched joist resting on flange of steel beam

crack propagates from notch in joist

joist

steel beam

mortise and tenon connection

crack propagates from bottom of tenon

joist

wood beam

notched joist resting on ledger board

crack propagates from notch in joist

joist

built-up wood beam

ledger board

0101

Watch for insufficient nails in joist hangers

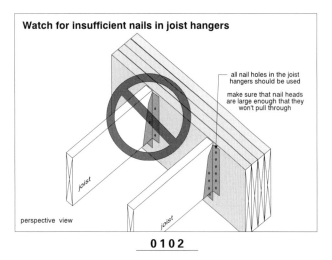

all nail holes in the joist hangers should be used

make sure that nail heads are large enough that they won't pull through

joist

joist

perspective view

0102

Lateral support for steel beams

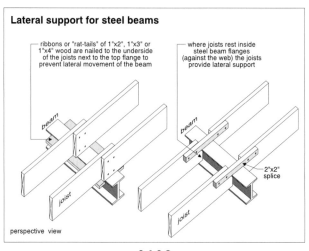

ribbons or "rat-tails" of 1"x2", 1"x3" or 1"x4" wood are nailed to the underside of the joists next to the top flange to prevent lateral movement of the beam

where joists rest inside steel beam flanges (against the web) the joists provide lateral support

beam

beam

joist

2"x2" splice

joist

perspective view

0103

Lateral support for wood beams

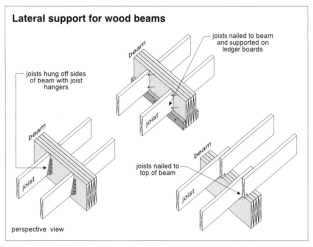

joists nailed to beam and supported on ledger boards

joists hung off sides of beam with joist hangers

beam

beam

joist

joist

joists nailed to top of beam

beam

joist

perspective view

0104

Concentrated loads - removing or altering walls

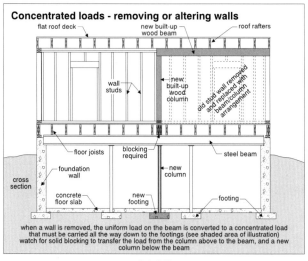

flat roof deck

new built-up wood beam

roof rafters

wall studs

new built-up wood column

old stud wall removed and replaced with beam/column arrangement

floor joists

blocking required

steel beam

foundation wall

new column

cross section

concrete floor slab

new footing

footing

when a wall is removed, the uniform load on the beam is converted to a concentrated load that must be carried all the way down to the footings (see shaded area of illustration) watch for solid blocking to transfer the load from the column above to the beam, and a new column below the beam

0105

Different types of floor joists

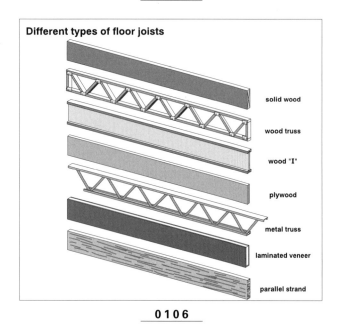

solid wood

wood truss

wood "I"

plywood

metal truss

laminated veneer

parallel strand

0106

Lateral support for masonry walls

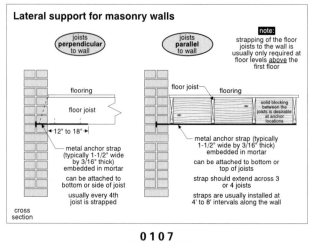

joists perpendicular to wall

joists parallel to wall

note:
strapping of the floor joists to the wall is usually only required at floor levels above the first floor

flooring

floor joist

floor joist

flooring

solid blocking between the joists is desirable at anchor locations

12" to 18"

metal anchor strap (typically 1-1/2" wide by 3/16" thick) embedded in mortar

can be attached to bottom or side of joist

usually every 4th joist is strapped

metal anchor strap (typically 1-1/2" wide by 3/16" thick) embedded in mortar

can be attached to bottom or top of joists

strap should extend across 3 or 4 joists

straps are usually installed at 4' to 8' intervals along the wall

cross section

0107

Fire cut joists
(with solid masonry exterior walls)

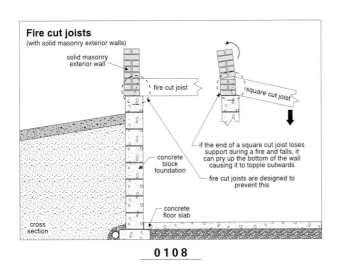

solid masonry exterior wall

fire cut joist

square cut joist

concrete block foundation

if the end of a square cut joist loses support during a fire and falls, it can pry up the bottom of the wall causing it to topple outwards

fire cut joists are designed to prevent this

concrete floor slab

cross section

0108

Joist installation - crown up versus crown down

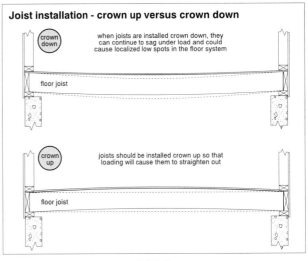

crown down

when joists are installed crown down, they can continue to sag under load and could cause localized low spots in the floor system

floor joist

crown up

joists should be installed crown up so that loading will cause them to straighten out

floor joist

0109

Rim joists
(also known as header or band joists)

rim joist is toe-nailed into the sill and end-nailed to the joists

the rim joist should be doubled over foundation wall openings such as windows

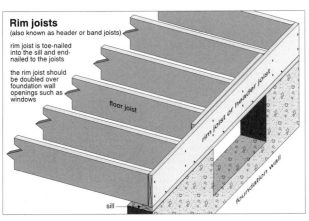

floor joist

rim joist or header joist

foundation wall

sill

0110

Two methods for improving sagging joists

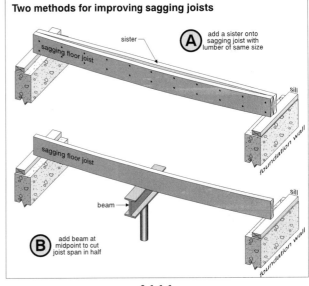

sagging floor joist

sister

(A) add a sister onto sagging joist with lumber of same size

sill

foundation wall

sagging floor joist

beam

(B) add beam at midpoint to cut joist span in half

sill

foundation wall

0111

Strapping, bridging and blocking

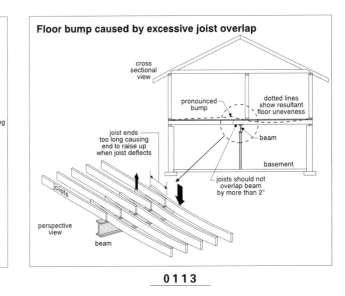

strapping

cross-bridging

solid blocking

joists

all of these methods are commonly used to reduce joist twisting and rotating when the ceilings below are not finished

another benefit of bridging and blocking is that load sharing between joists and vibration damping are also improved

perspective

0112

Floor bump caused by excessive joist overlap

cross sectional view

pronounced bump

dotted lines show resultant floor uneveness

joist ends too long causing end to raise up when joist deflects

beam

basement

joists should not overlap beam by more than 2"

joists

perspective view

beam

0113

Different types of joist end support

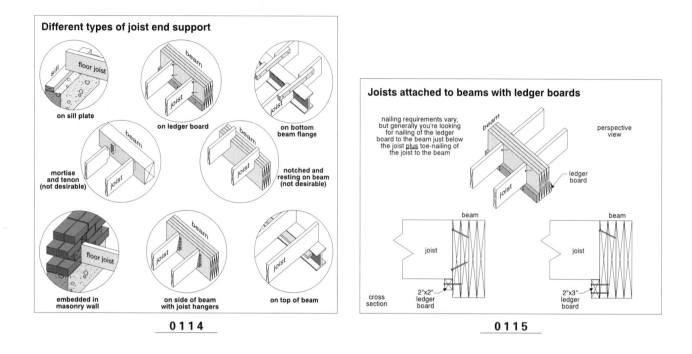

on sill plate

on ledger board

on bottom beam flange

mortise and tenon (not desirable)

notched and resting on beam (not desirable)

embedded in masonry wall

on side of beam with joist hangers

on top of beam

0114

Joists attached to beams with ledger boards

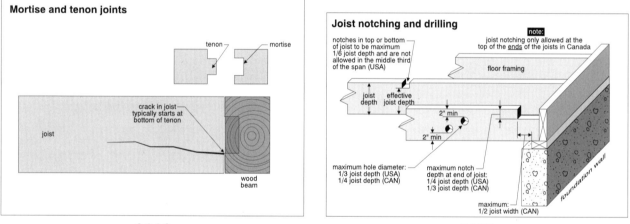

nailing requirements vary, but generally you're looking for nailing of the ledger board to the beam just below the joist plus toe-nailing of the joist to the beam

perspective view

ledger board

cross section

2"x2" ledger board

2"x3" ledger board

0115

Mortise and tenon joints

tenon — mortise

joist

crack in joist typically starts at bottom of tenon

wood beam

0116

Joist notching and drilling

note:
joist notching only allowed at the top of the ends of the joists in Canada

notches in top or bottom of joist to be maximum 1/6 joist depth and are not allowed in the middle third of the span (USA)

floor framing

joist depth   effective joist depth

2" min

2" min

foundation wall

maximum hole diameter:
1/3 joist depth (USA)
1/4 joist depth (CAN)

maximum notch depth at end of joist:
1/4 joist depth (USA)
1/3 joist depth (CAN)

maximum:
1/2 joist width (CAN)

0117

Common causes of cracked joists

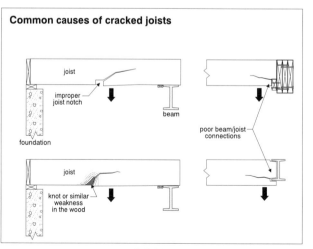

joist

improper joist notch

beam

foundation

poor beam/joist connections

joist

knot or similar weakness in the wood

0118

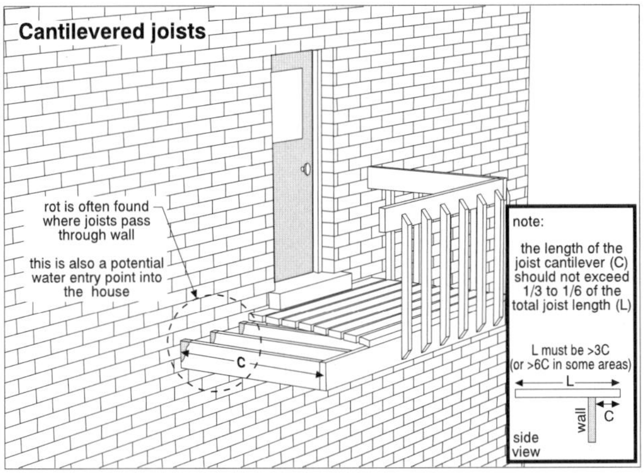

Cantilevered joists

rot is often found where joists pass through wall

this is also a potential water entry point into the house

C

note:

the length of the joist cantilever (C) should not exceed 1/3 to 1/6 of the total joist length (L)

L must be >3C (or >6C in some areas)

L

wall

C

side view

0119

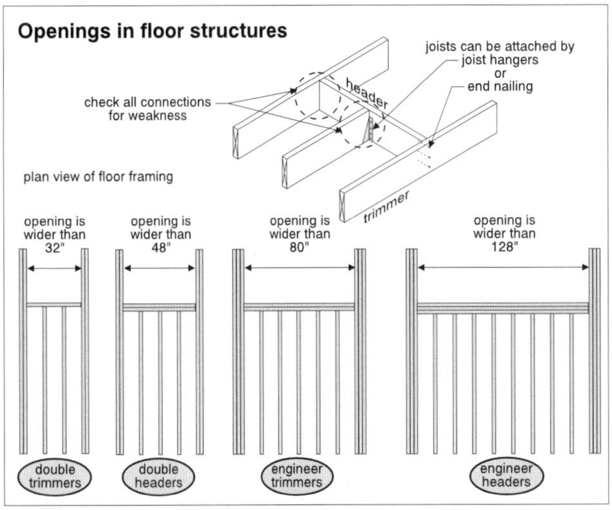

Openings in floor structures

joists can be attached by joist hangers
or
end nailing

header

check all connections for weakness

plan view of floor framing

opening is wider than 32"	opening is wider than 48"	opening is wider than 80"	opening is wider than 128"
double trimmers	double headers	engineer trimmers	engineer headers

trimmer

0120

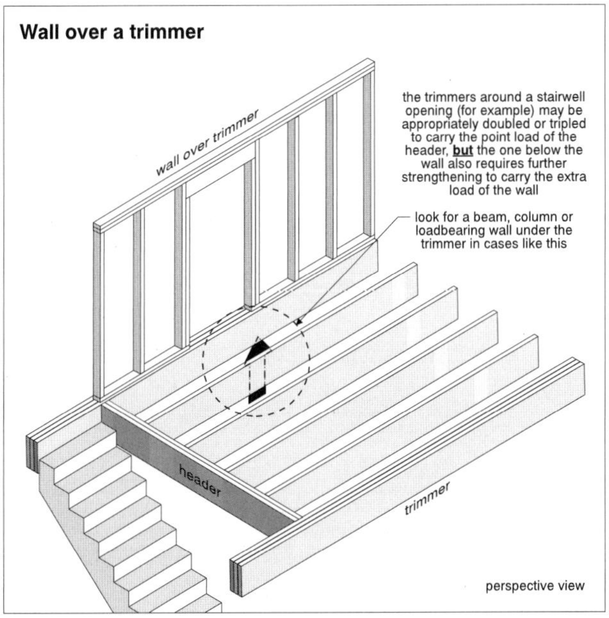

Wall over a trimmer

wall over trimmer

the trimmers around a stairwell opening (for example) may be appropriately doubled or tripled to carry the point load of the header, **but** the one below the wall also requires further strengthening to carry the extra load of the wall

look for a beam, column or loadbearing wall under the trimmer in cases like this

header

trimmer

perspective view

0121

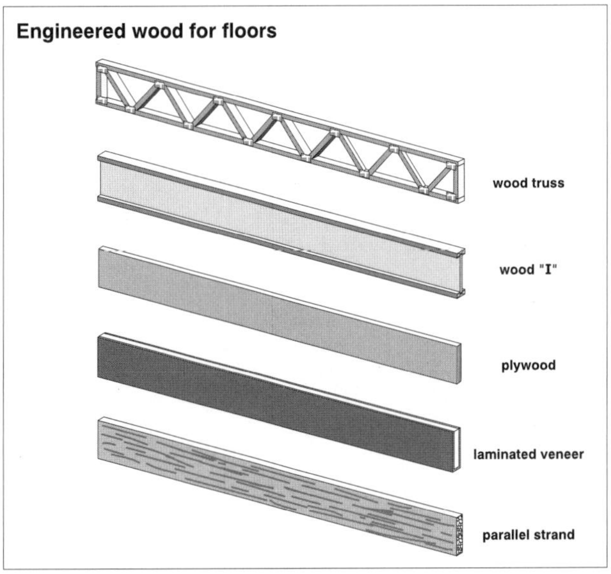

Engineered wood for floors

wood truss

wood "I"

plywood

laminated veneer

parallel strand

0122

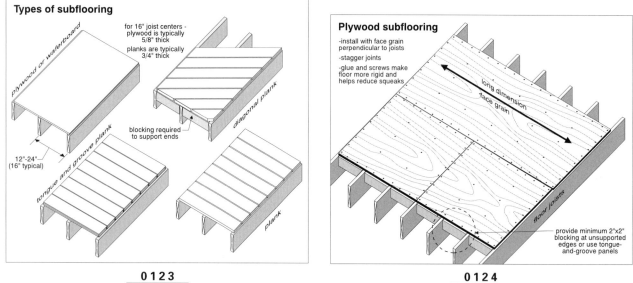

Types of subflooring

plywood or waferboard

for 16" joist centers - plywood is typically 5/8" thick

planks are typically 3/4" thick

diagonal plank

12"-24" (16" typical)

tongue and groove plank

blocking required to support ends

plank

0123

Plywood subflooring

-install with face grain perpendicular to joists

-stagger joints

-glue and screws make floor more rigid and helps reduce squeaks

long dimension face grain

floor joists

provide minimum 2"x2" blocking at unsupported edges or use tongue-and-groove panels

0124

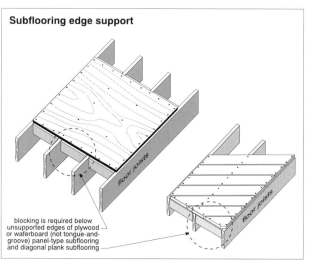

Subflooring edge support

floor joists

floor joists

blocking is required below unsupported edges of plywood or waferboard (not tongue-and-groove) panel-type subflooring and diagonal plank subflooring

0125

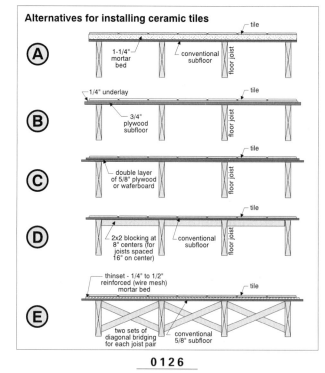

Alternatives for installing ceramic tiles

(A)   tile
1-1/4" mortar bed   conventional subfloor   floor joist

(B)   1/4" underlay   tile
3/4" plywood subfloor   floor joist

(C)   tile
double layer of 5/8" plywood or waferboard   floor joist

(D)   tile
2x2 blocking at 8" centers (for joists spaced 16" on center)   conventional subfloor   floor joist

(E)   thinset - 1/4" to 1/2" reinforced (wire mesh) mortar bed   tile
two sets of diagonal bridging for each joist pair   conventional 5/8" subfloor

0126

Prestressed and post-tensioned concrete floor slabs

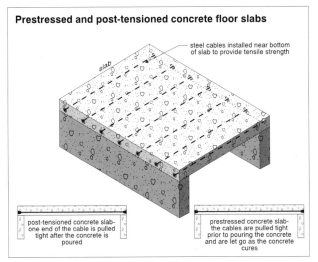

steel cables installed near bottom of slab to provide tensile strength

slab

post-tensioned concrete slab- one end of the cable is pulled tight after the concrete is poured

prestressed concrete slab- the cables are pulled tight prior to pouring the concrete and are let go as the concrete cures

0 1 2 7

Cold joints along edge of slabs

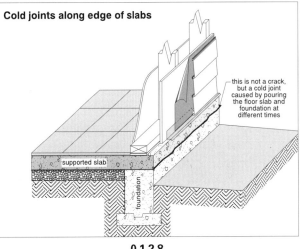

this is not a crack, but a cold joint caused by pouring the floor slab and foundation at different times

supported slab

foundation

0 1 2 8

Solid masonry walls

cross sectional view

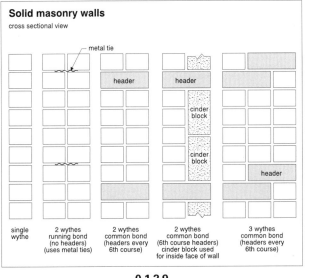

metal tie

header

header

cinder block

cinder block

header

single wythe

2 wythes running bond (no headers) (uses metal ties)

2 wythes common bond (headers every 6th course)

2 wythes common bond (6th course headers) cinder block used for inside face of wall

3 wythes common bond (headers every 6th course)

0 1 2 9

Masonry cavity wall

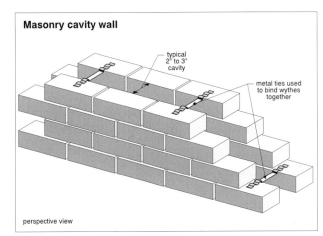

typical 2" to 3" cavity

metal ties used to bind wythes together

perspective view

0 1 3 0

Brick veneer wall

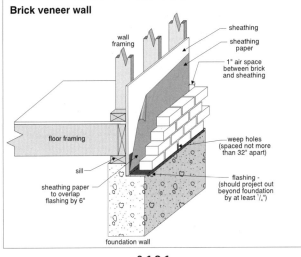

wall framing

sheathing

sheathing paper

1" air space between brick and sheathing

floor framing

sill

sheathing paper to overlap flashing by 6"

weep holes (spaced not more than 32" apart)

flashing - (should project out beyond foundation by at least 1/4")

foundation wall

0 1 3 1

Brick wall terminology

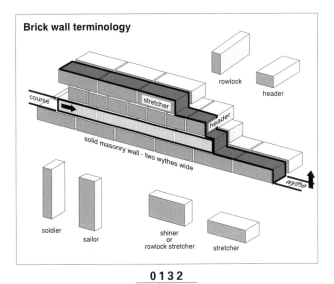

rowlock

header

course

stretcher

header

solid masonry wall - two wythes wide

wythe

soldier

sailor

shiner or rowlock stretcher

stretcher

0 1 3 2

Masonry wall using a diagonal brick bond

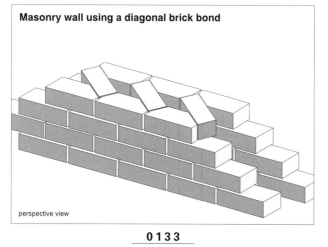

perspective view

0 1 3 3

Lateral support for exterior masonry walls

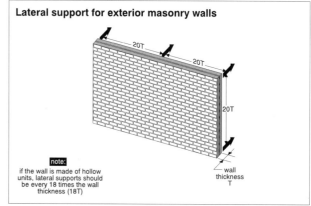

20T
20T
20T
wall thickness T

note:
if the wall is made of hollow units, lateral supports should be every 18 times the wall thickness (18T)

0 1 3 4

Walls that extend above ceiling joists

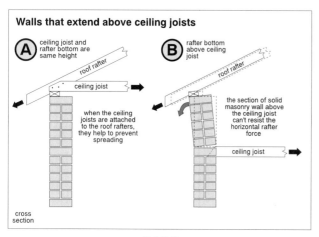

(A) ceiling joist and rafter bottom are same height

roof rafter
ceiling joist

when the ceiling joists are attached to the roof rafters, they help to prevent spreading

(B) rafter bottom above ceiling joist

roof rafter

the section of solid masonry wall above the ceiling joist can't resist the horizontal rafter force

ceiling joist

cross section

0 1 3 5

Steel lintel in brick veneer wall

cutaway view

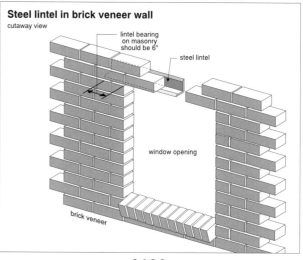

lintel bearing on masonry should be 6"

steel lintel

window opening

brick veneer

0 1 3 6

Wood lintel in solid masonry wall

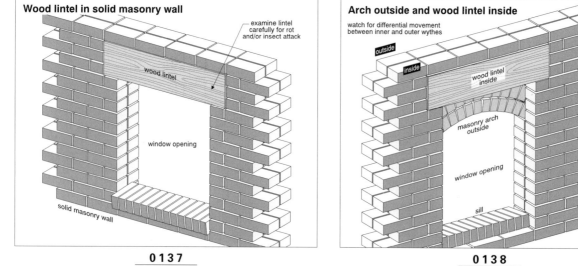

examine lintel carefully for rot and/or insect attack

wood lintel

window opening

solid masonry wall

0 1 3 7

Arch outside and wood lintel inside

watch for differential movement between inner and outer wythes

outside
inside

wood lintel inside

masonry arch outside

window opening

sill

0 1 3 8

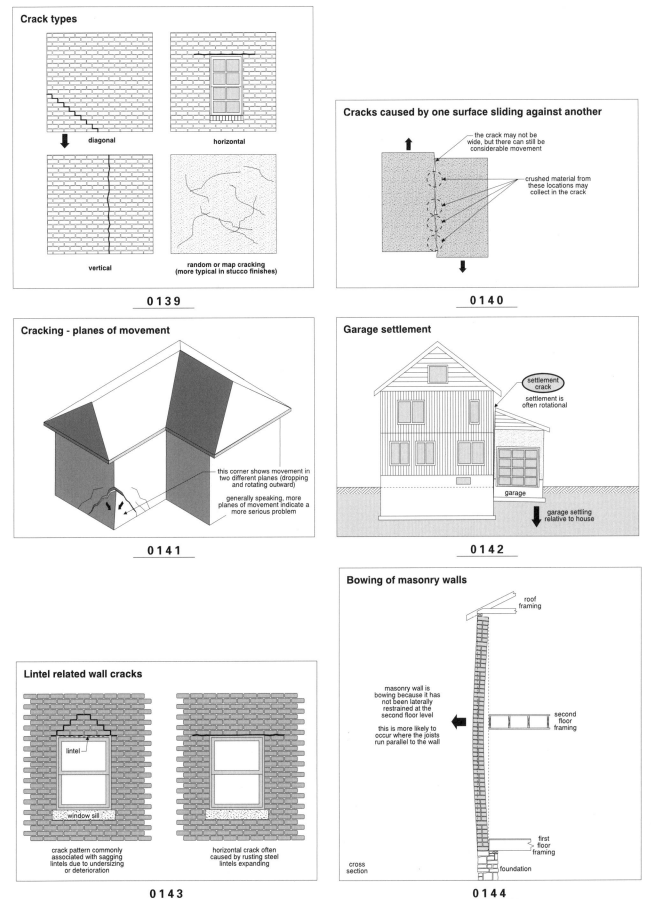

Crack types

diagonal

horizontal

vertical

random or map cracking
(more typical in stucco finishes)

0139

Cracks caused by one surface sliding against another

the crack may not be
wide, but there can still be
considerable movement

crushed material from
these locations may
collect in the crack

0140

Cracking - planes of movement

this corner shows movement in
two different planes (dropping
and rotating outward)

generally speaking, more
planes of movement indicate a
more serious problem

0141

Garage settlement

settlement
crack

settlement is
often rotational

garage

garage settling
relative to house

0142

Lintel related wall cracks

lintel

window sill

crack pattern commonly
associated with sagging
lintels due to undersizing
or deterioration

horizontal crack often
caused by rusting steel
lintels expanding

0143

Bowing of masonry walls

roof
framing

masonry wall is
bowing because it has
not been laterally
restrained at the
second floor level

this is more likely to
occur where the joists
run parallel to the wall

second
floor
framing

first
floor
framing

cross
section

foundation

0144

Bowed brick veneer wall - older home

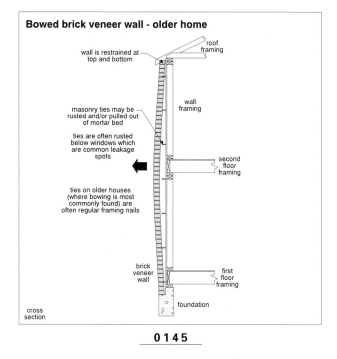

wall is restrained at top and bottom

roof framing

wall framing

masonry ties may be rusted and/or pulled out of mortar bed

ties are often rusted below windows which are common leakage spots

second floor framing

ties on older houses (where bowing is most commonly found) are often regular framing nails

brick veneer wall

first floor framing

foundation

cross section

0145

Bowing of solid masonry wall due to tie failure

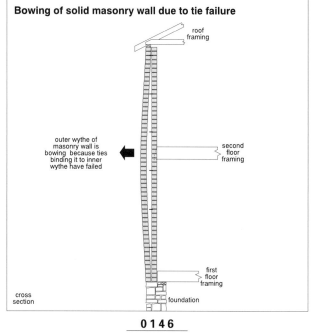

roof framing

outer wythe of masonry wall is bowing because ties binding it to inner wythe have failed

second floor framing

first floor framing

foundation

cross section

0146

Cracks due to clay brick expansion

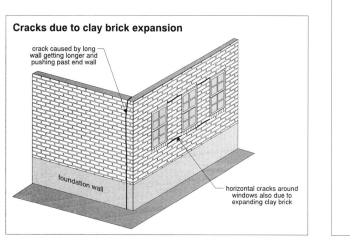

crack caused by long wall getting longer and pushing past end wall

foundation wall

horizontal cracks around windows also due to expanding clay brick

0147

Corbelling

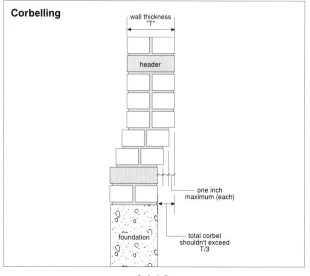

wall thickness "T"

header

one inch maximum (each)

foundation

total corbel shouldn't exceed T/3

0148

Projections for veneer

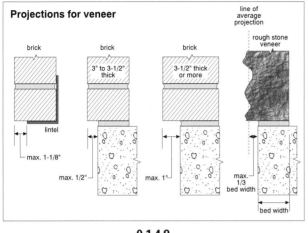

brick

brick
3" to 3-1/2" thick

brick
3-1/2" thick or more

line of average projection

rough stone veneer

lintel

max. 1-1/8"

max. 1/2"

max. 1"

max. 1/3 bed width

bed width

0149

Blocks and bricks - hollow channels must be vertical

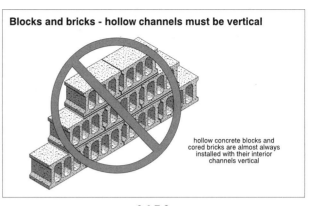

hollow concrete blocks and cored bricks are almost always installed with their interior channels vertical

0150

Platform versus balloon framing

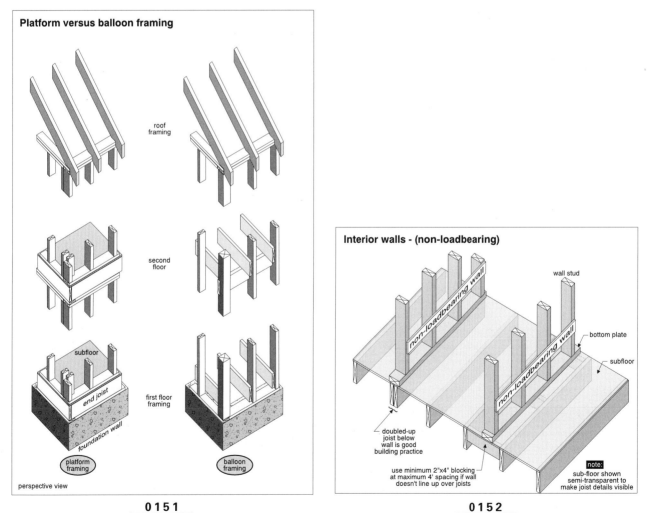

roof framing

second floor

first floor framing

subfloor

end joist

foundation wall

platform framing

balloon framing

perspective view

0151

Interior walls - (non-loadbearing)

non-loadbearing wall

non-loadbearing wall

wall stud

bottom plate

subfloor

doubled-up joist below wall is good building practice

use minimum 2"x4" blocking at maximum 4' spacing if wall doesn't line up over joists

note: sub-floor shown semi-transparent to make joist details visible

0152

Wood frame bearing wall - top plate T-connection

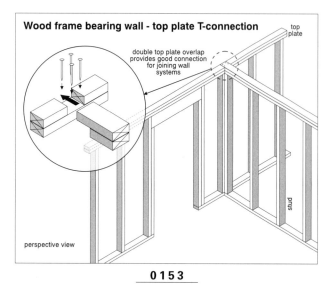

double top plate overlap provides good connection for joining wall systems

top plate

stud

perspective view

0153

Holes and notches in top plates of bearing walls

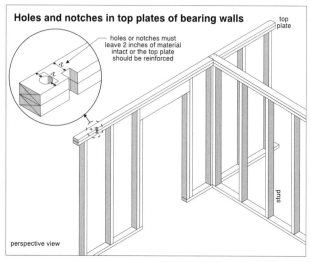

holes or notches must leave 2 inches of material intact or the top plate should be reinforced

top plate

stud

perspective view

0154

Metal ties for connections in single top plates

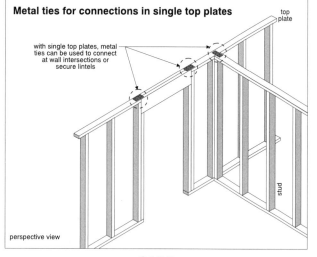

with single top plates, metal ties can be used to connect at wall intersections or secure lintels

top plate

stud

perspective view

0155

Common rot locations in walls

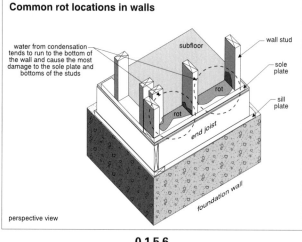

water from condensation tends to run to the bottom of the wall and cause the most damage to the sole plate and bottoms of the studs

subfloor

wall stud

sole plate

sill plate

rot

rot

end joist

foundation wall

perspective view

0156

Wood/soil clearances

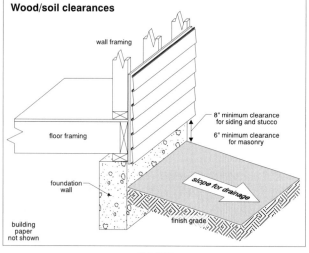

wall framing

floor framing

foundation wall

building paper not shown

8" minimum clearance for siding and stucco

6" minimum clearance for masonry

slope for drainage

finish grade

0157

Wood frame wall - racking resistance

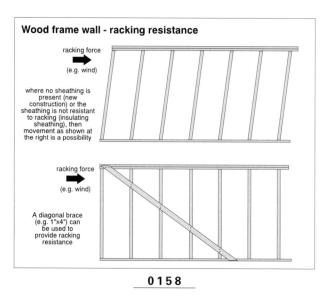

racking force
(e.g. wind)

where no sheathing is present (new construction) or the sheathing is not resistant to racking (insulating sheathing), then movement as shown at the right is a possibility

racking force
(e.g. wind)

A diagonal brace (e.g. 1"x4") can be used to provide racking resistance

0158

Wood frame bearing wall - in basement

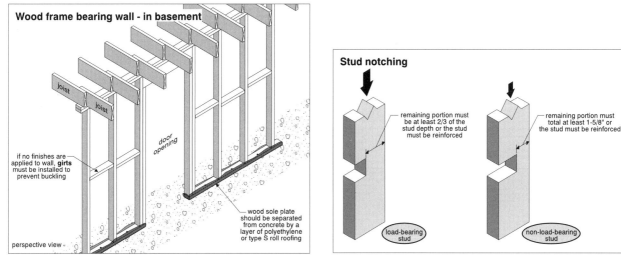

joist

joist

door opening

if no finishes are applied to wall, **girts** must be installed to prevent buckling

wood sole plate should be separated from concrete by a layer of polyethylene or type S roll roofing

perspective view

0159

Stud notching

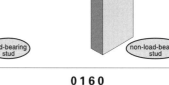

remaining portion must be at least 2/3 of the stud depth or the stud must be reinforced

load-bearing stud

remaining portion must total at least 1-5/8" or the stud must be reinforced

non-load-bearing stud

0160

Straightening partition studs

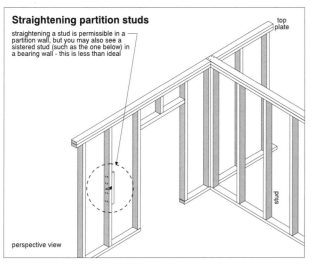

straightening a stud is permissible in a partition wall, but you may also see a sistered stud (such as the one below) in a bearing wall - this is less than ideal

top plate

stud

perspective view

0161

Concentrated loads - removing or altering walls

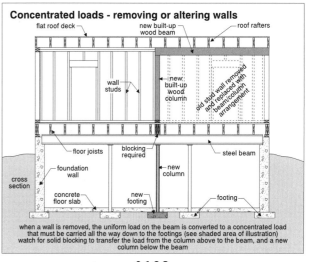

flat roof deck

new built-up wood beam

roof rafters

wall studs

new built-up wood column

old stud wall removed and replaced with beam/column arrangement

floor joists

blocking required

new column

steel beam

foundation wall

concrete floor slab

new footing

footing

cross section

when a wall is removed, the uniform load on the beam is converted to a concentrated load that must be carried all the way down to the footings (see shaded area of illustration) watch for solid blocking to transfer the load from the column above to the beam, and a new column below the beam

0162

Support for jack studs on wide openings

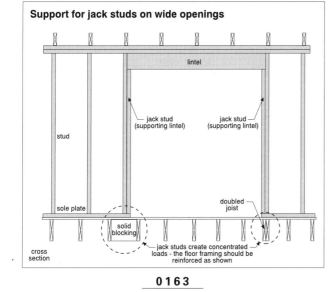

lintel

jack stud (supporting lintel)

jack stud (supporting lintel)

stud

sole plate

doubled joist

solid blocking

jack studs create concentrated loads - the floor framing should be reinforced as shown

cross section

0163

Offset bearing walls

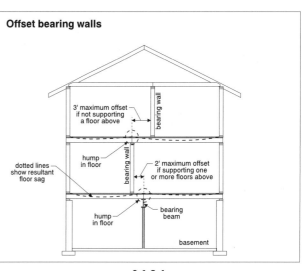

bearing wall

3' maximum offset if not supporting a floor above

hump in floor

bearing wall

dotted lines show resultant floor sag

2' maximum offset if supporting one or more floors above

hump in floor

bearing beam

basement

0164

Sagging interior lintel

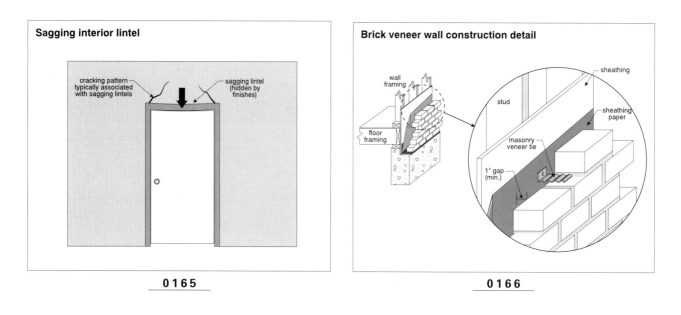

cracking pattern typically associated with sagging lintels

sagging lintel (hidden by finishes)

0165

Brick veneer wall construction detail

wall framing

floor framing

sheathing

stud

sheathing paper

masonry veneer tie

1" gap (min.)

0166

Brick veneer wall - masonry tie detail

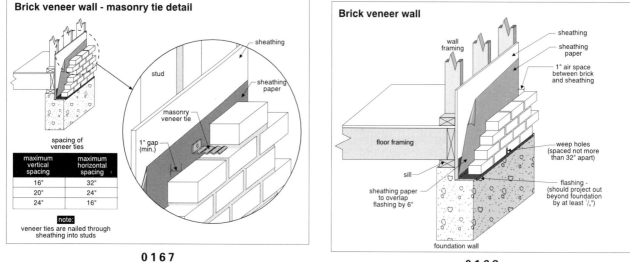

sheathing

stud

sheathing paper

masonry veneer tie

1" gap (min.)

spacing of veneer ties

maximum vertical spacing	maximum horizontal spacing
16"	32"
20"	24"
24"	16"

note:
veneer ties are nailed through sheathing into studs

0167

Brick veneer wall

wall framing

floor framing

sill

sheathing paper to overlap flashing by 6"

foundation wall

sheathing

sheathing paper

1" air space between brick and sheathing

weep holes (spaced not more than 32" apart)

flashing - (should project out beyond foundation by at least 1/4")

0168

Weep holes - vented rain screen principle

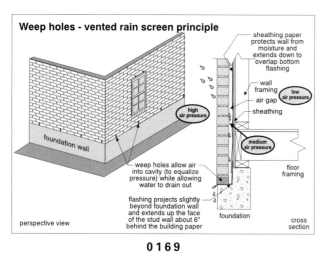

sheathing paper protects wall from moisture and extends down to overlap bottom flashing

wall framing

air gap

sheathing

high air pressure

low air pressure

medium air pressure

floor framing

foundation wall

weep holes allow air into cavity (to equalize pressure) while allowing water to drain out

flashing projects slightly beyond foundation wall and extends up the face of the stud wall about 6" behind the building paper

foundation

perspective view

cross section

0169

Weep holes in brick veneer walls

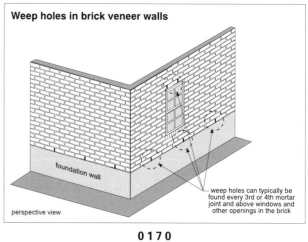

foundation wall

weep holes can typically be found every 3rd or 4th mortar joint and above windows and other openings in the brick

perspective view

0170

Veneer versus solid masonry

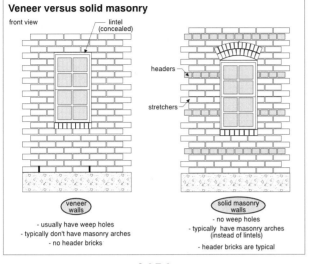

front view

lintel (concealed)

headers

stretchers

veneer walls

- usually have weep holes
- typically don't have masonry arches
- no header bricks

solid masonry walls

- no weep holes
- typically have masonry arches (instead of lintels)
- header bricks are typical

0171

Lintel loads in a masonry wall

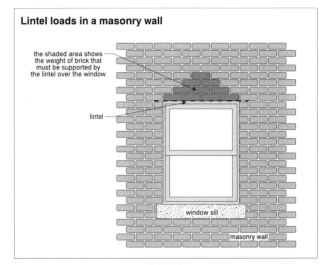

the shaded area shows the weight of brick that must be supported by the lintel over the window

lintel

window sill

masonry wall

0172

Lintel loads in wood frame walls

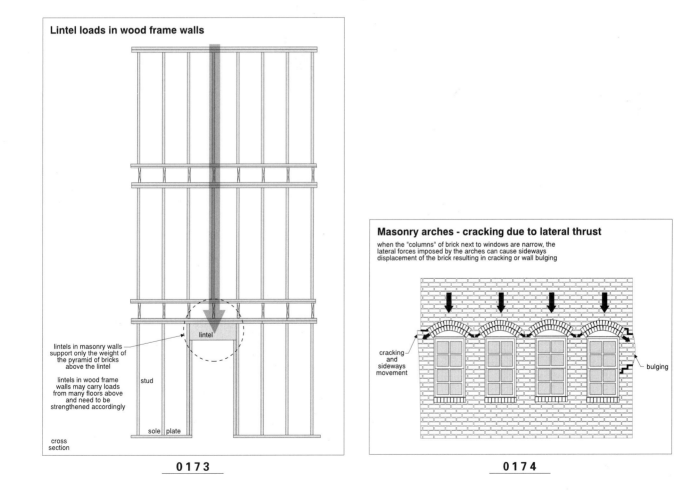

lintel

lintels in masonry walls support only the weight of the pyramid of bricks above the lintel

lintels in wood frame walls may carry loads from many floors above and need to be strengthened accordingly

stud

sole plate

cross section

0173

Masonry arches - cracking due to lateral thrust

when the "columns" of brick next to windows are narrow, the lateral forces imposed by the arches can cause sideways displacement of the brick resulting in cracking or wall bulging

cracking and sideways movement

bulging

0174

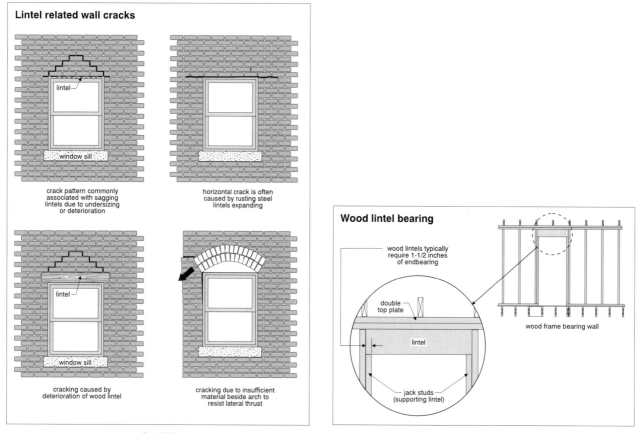

Lintel related wall cracks

lintel

window sill

crack pattern commonly associated with sagging lintels due to undersizing or deterioration

horizontal crack is often caused by rusting steel lintels expanding

lintel

window sill

cracking caused by deterioration of wood lintel

cracking due to insufficient material beside arch to resist lateral thrust

0175

Wood lintel bearing

wood lintels typically require 1-1/2 inches of endbearing

double top plate

lintel

jack studs (supporting lintel)

wood frame bearing wall

0176

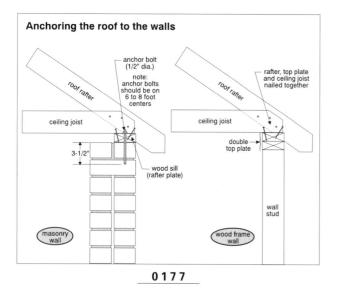

Anchoring the roof to the walls

anchor bolt (1/2" dia.)

note: anchor bolts should be on 6 to 8 foot centers

roof rafter

ceiling joist

3-1/2"

wood sill (rafter plate)

masonry wall

rafter, top plate and ceiling joist nailed together

roof rafter

ceiling joist

double top plate

wall stud

wood frame wall

0177

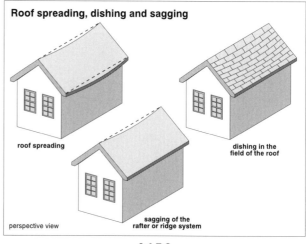

Roof spreading, dishing and sagging

roof spreading

dishing in the field of the roof

sagging of the rafter or ridge system

perspective view

0178

Roof joists versus roof rafters

when the roof slope is 2 in 12 or less, the
primary roof supports are called roof joists
on steeper roofs, they are called roof rafters

12
2

roof
rafter

12
>2

roof
joist

ceiling joist

ceiling joist
(may or may not
be present)

wall
stud

wall
stud

0179

Overlapped ceiling joist splices

rafter

ceiling
joist

a good connection is needed
at ceiling joist splices so the
ends of opposing rafters are
adequately restrained

the most common connection
method is to overlap the joists
over a central bearing wall and
nail them together

top
plate

wall studs

perspective view

0180

Plywood ceiling joist splices

rafter

ceiling
joist

a good connection is needed at
ceiling joist splices so the ends
of opposing rafters are
adequately restrained

a less common, but high quality,
connection is to nail a 2' to 3'
wide piece of plywood to the top
of the joists (this provides a good
connection and helps to better
distribute the load)

top
plate

wall studs

perspective view

0181

Rafter endbearing

roof rafter

rafter
plate

typical endbearing
for rafters, roof
joists and ceiling
joists is 1-1/2"

toe bearing is poor
framing practice and
can result in cracked
roof members

bird's
mouth

0182

Hip and valley rafters

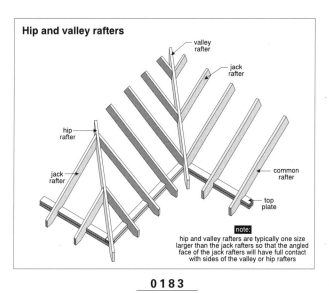

valley
rafter

jack
rafter

jack
rafter

hip
rafter

common
rafter

top
plate

note:
hip and valley rafters are typically one size
larger than the jack rafters so that the angled
face of the jack rafters will have full contact
with sides of the valley or hip rafters

0183

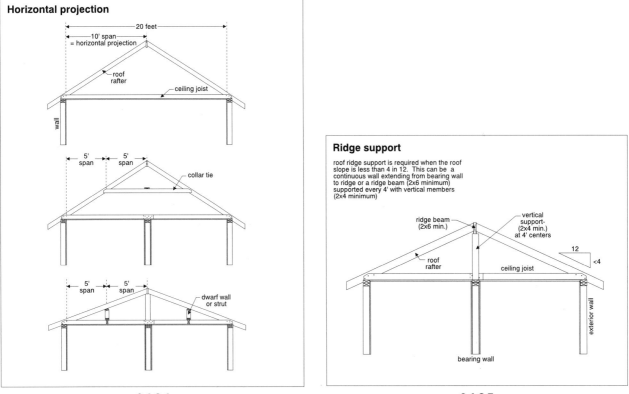

Horizontal projection

20 feet

10' span
= horizontal projection

roof rafter

ceiling joist

wall

5' span   5' span

collar tie

5' span   5' span

dwarf wall or strut

0184

Ridge support

roof ridge support is required when the roof slope is less than 4 in 12. This can be a continuous wall extending from bearing wall to ridge or a ridge beam (2x6 minimum) supported every 4' with vertical members (2x4 minimum)

ridge beam (2x6 min.)

vertical support- (2x4 min.) at 4' centers

roof rafter

ceiling joist

12
<4

exterior wall

bearing wall

0185

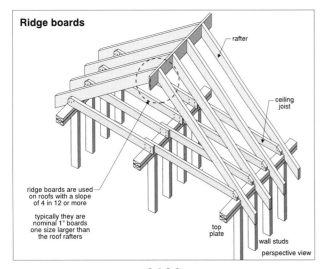

Ridge boards

rafter

ceiling joist

ridge boards are used on roofs with a slope of 4 in 12 or more

typically they are nominal 1" boards one size larger than the roof rafters

top plate

wall studs

perspective view

0186

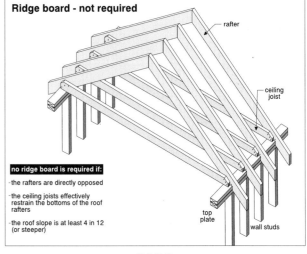

Ridge board - not required

rafter

ceiling joist

no ridge board is required if:

- the rafters are directly opposed

- the ceiling joists effectively restrain the bottoms of the roof rafters

- the roof slope is at least 4 in 12 (or steeper)

top plate

wall studs

0187

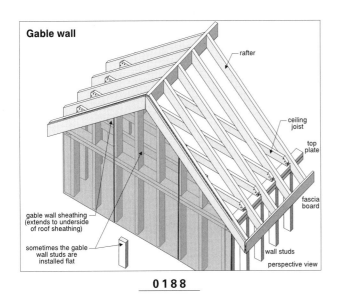

Gable wall

rafter

ceiling joist

top plate

gable wall sheathing (extends to underside of roof sheathing)

sometimes the gable wall studs are installed flat

fascia board

wall studs

perspective view

0188

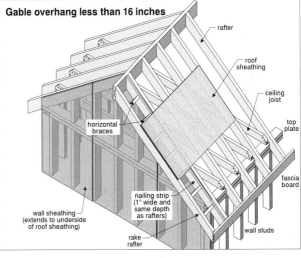

Gable overhang less than 16 inches

rafter

roof sheathing

ceiling joist

top plate

horizontal braces

nailing strip (1" wide and same depth as rafters)

fascia board

wall sheathing (extends to underside of roof sheathing)

rake rafter

wall studs

0189

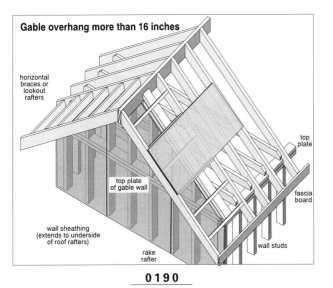

Gable overhang more than 16 inches

horizontal braces or lookout rafters

top plate

top plate of gable wall

fascia board

wall sheathing (extends to underside of roof rafters)

rake rafter

wall studs

0190

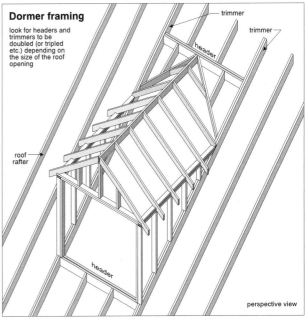

Dormer framing

look for headers and trimmers to be doubled (or tripled etc.) depending on the size of the roof opening

trimmer

trimmer

header

roof rafter

header

perspective view

0191

Soffits and fascia

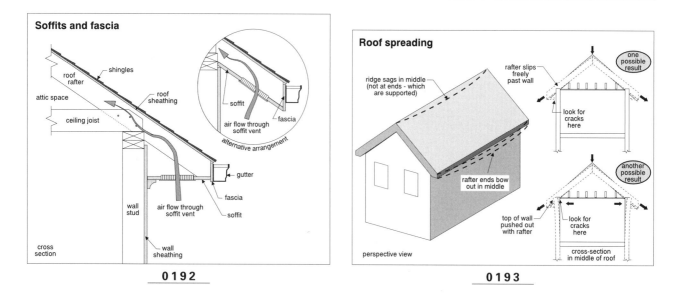

shingles

roof rafter

attic space

roof sheathing

ceiling joist

soffit

air flow through soffit vent

alternative arrangement

fascia

gutter

fascia

soffit

wall stud

air flow through soffit vent

wall sheathing

cross section

0192

Roof spreading

ridge sags in middle (not at ends - which are supported)

rafter ends bow out in middle

rafter slips freely past wall

one possible result

look for cracks here

another possible result

top of wall pushed out with rafter

look for cracks here

cross-section in middle of roof

perspective view

0193

Roof spreading - remedial action

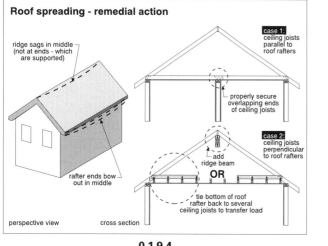

ridge sags in middle (not at ends - which are supported)

rafter ends bow out in middle

case 1: ceiling joists parallel to roof rafters

properly secure overlapping ends of ceiling joists

case 2: ceiling joists perpendicular to roof rafters

add ridge beam

OR

tie bottom of roof rafter back to several ceiling joists to transfer load

perspective view

cross section

0194

Watch for insufficient nails in joist hangers

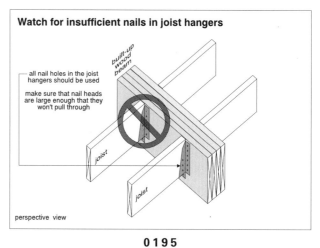

built-up wood beam

all nail holes in the joist hangers should be used

make sure that nail heads are large enough that they won't pull through

joist

joist

perspective view

0195

Openings in roofs

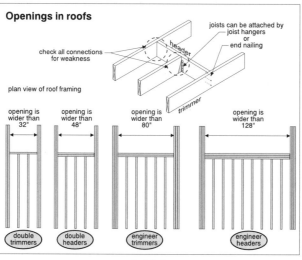

check all connections for weakness

header

joists can be attached by joist hangers or end nailing

trimmer

plan view of roof framing

opening is wider than 32"

opening is wider than 48"

opening is wider than 80"

opening is wider than 128"

double trimmers

double headers

engineer trimmers

engineer headers

0196

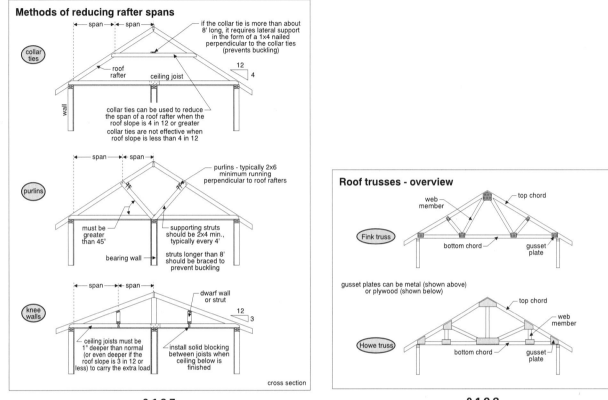

Methods of reducing rafter spans

collar ties

|← span →|← span →|

if the collar tie is more than about 8' long, it requires lateral support in the form of a 1x4 nailed perpendicular to the collar ties (prevents buckling)

roof rafter

ceiling joist

12 / 4

wall

collar ties can be used to reduce the span of a roof rafter when the roof slope is 4 in 12 or greater

collar ties are not effective when roof slope is less than 4 in 12

purlins

|← span →|← span →|

purlins - typically 2x6 minimum running perpendicular to roof rafters

must be greater than 45°

supporting struts should be 2x4 min., typically every 4'

bearing wall

struts longer than 8' should be braced to prevent buckling

knee walls

|← span →|← span →|

dwarf wall or strut

12 / 3

ceiling joists must be 1" deeper than normal (or even deeper if the roof slope is 3 in 12 or less) to carry the extra load

install solid blocking between joists when ceiling below is finished

cross section

0197

Roof trusses - overview

web member

top chord

Fink truss

bottom chord

gusset plate

gusset plates can be metal (shown above) or plywood (shown below)

top chord

web member

Howe truss

bottom chord

gusset plate

0198

Bracing of compression webs

long compression webs may require bracing (typically done with 1x4's) near the midpoints

wall

wall

0199

Strapping the underside of trusses

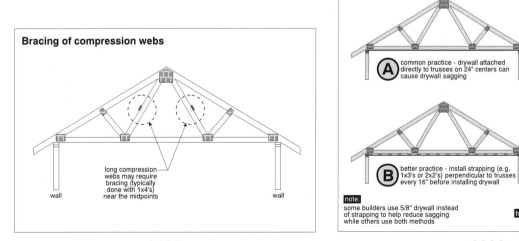

(A) common practice - drywall attached directly to trusses on 24" centers can cause drywall sagging

(B) better practice - install strapping (e.g. 1x3's or 2x2's) perpendicular to trusses every 16" before installing drywall

note:
some builders use 5/8" drywall instead of strapping to help reduce sagging while others use both methods

trusses

sagging drywall

trusses

drywall applied to strapping

front view side view

0200

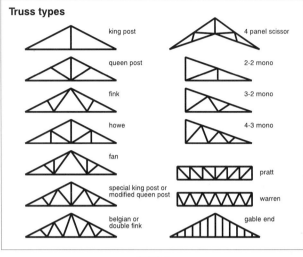

Truss types

king post

queen post

fink

howe

fan

special king post or
modified queen post

belgian or
double fink

4 panel scissor

2-2 mono

3-2 mono

4-3 mono

pratt

warren

gable end

0201

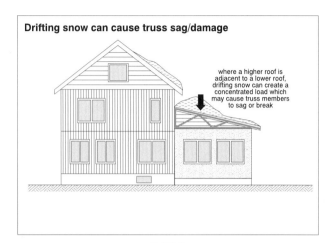

Drifting snow can cause truss sag/damage

where a higher roof is
adjacent to a lower roof,
drifting snow can create a
concentrated load which
may cause truss members
to sag or break

0202

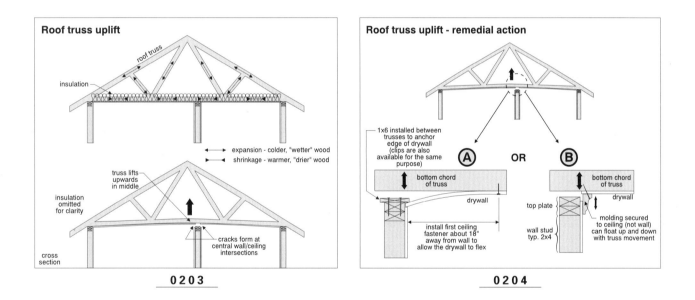

Roof truss uplift

roof truss

insulation

expansion - colder, "wetter" wood

shrinkage - warmer, "drier" wood

truss lifts
upwards
in middle

insulation
omitted
for clarity

cracks form at
central wall/ceiling
intersections

cross
section

0203

Roof truss uplift - remedial action

1x6 installed between
trusses to anchor
edge of drywall
(clips are also
available for the same
purpose)

Ⓐ OR Ⓑ

bottom chord
of truss

bottom chord
of truss

drywall

drywall

install first ceiling
fastener about 18"
away from wall to
allow the drywall to flex

top plate

wall stud
typ. 2x4

molding secured
to ceiling (not wall)
can float up and down
with truss movement

0204

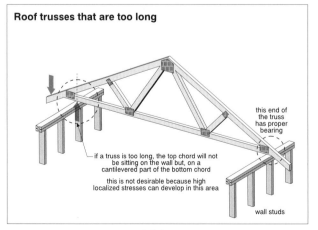

Roof trusses that are too long

this end of
the truss
has proper
bearing

if a truss is too long, the top chord will not
be sitting on the wall but, on a
cantilevered part of the bottom chord

this is not desirable because high
localized stresses can develop in this area

wall studs

0205

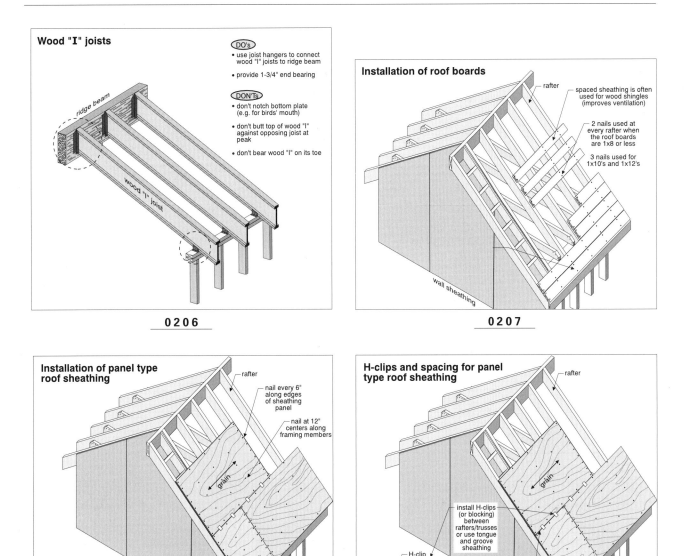

Wood "I" joists

DO's
- use joist hangers to connect wood "I" joists to ridge beam
- provide 1-3/4" end bearing

DON'Ts
- don't notch bottom plate (e.g. for birds' mouth)
- don't butt top of wood "I" against opposing joist at peak
- don't bear wood "I" on its toe

ridge beam

wood "I" joist

0206

Installation of roof boards

rafter

spaced sheathing is often used for wood shingles (improves ventilation)

2 nails used at every rafter when the roof boards are 1x8 or less

3 nails used for 1x10's and 1x12's

wall sheathing

0207

Installation of panel type roof sheathing

rafter

nail every 6" along edges of sheathing panel

nail at 12" centers along framing members

grain

wall sheathing

0208

H-clips and spacing for panel type roof sheathing

rafter

grain

install H-clips (or blocking) between rafters/trusses or use tongue and groove sheathing

provide 1/8" gap

wall sheathing

H-clip

cross section

sheathing

0209

PART 2

ROOFING

FLAT ROOF FLASHINGS

0305 Built-up flat roof/wall flashing
0306 Modified bitumen roof/wall flashing
0307 Two-piece counterflashings
0308 Built-up roofing membrane – edge details
0309 Built-up roof membrane – curbed edge
0310 Alternative attachment method for counter flashing
0311 Lower quality built-up flat roof/wall flashing
0312 EPDM (or PVC) roof/wall flashing

0313 Through-wall flashings
0314 Metal clad parapet wall
0315 Chimney cap details
0316 Skylight in flat roof
0317 "Stack Jack" plumbing stack flashing
0318 Cone-style plumbing stack flashing
0319 Pitch pans for irregular roof penetrations
0320 Drain in modified bitumen roof
0321 Flat roof draining onto sloped roof
0322 Sloped roof draining onto flat roof

Roof slopes

slope=rise/run

conventional slope   (4 in 12 and up)

4 (rise)

12 (run)

to   4°   12

low slope   (2 in 12 to 4 in 12)

2   12   to   4   12

flat   (0 in 12 to 2 in 12)

flat   to   2   12

0 2 1 0

Steep roof types

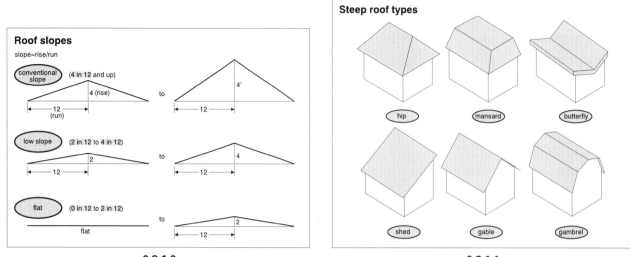

hip   mansard   butterfly

shed   gable   gambrel

0 2 1 1

Dormer gutters - discharging onto roof

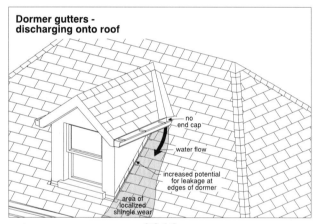

no end cap

water flow

increased potential for leakage at edges of dormer

area of localized shingle wear

0 2 1 2

Patched roofing

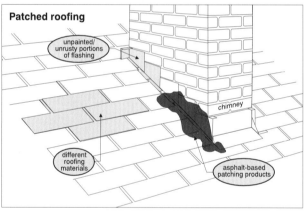

unpainted/ unrusty portions of flashing

chimney

different roofing materials

asphalt-based patching products

0 2 1 3

Vulnerable areas

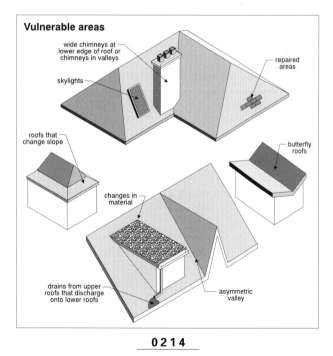

wide chimneys at lower edge of roof or chimneys in valleys

skylights

repaired areas

roofs that change slope

butterfly roofs

changes in material

drains from upper roofs that discharge onto lower roofs

asymmetric valley

0 2 1 4

Ice dams

clues to look for:

winter   summer

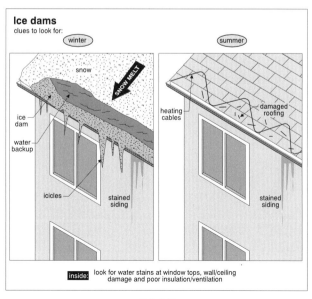

snow

SNOW MELT

ice dam

water backup

icicles   stained siding

heating cables

damaged roofing

stained siding

inside: look for water stains at window tops, wall/ceiling damage and poor insulation/ventilation

0 2 1 5

Preventing ice dams with ventilation

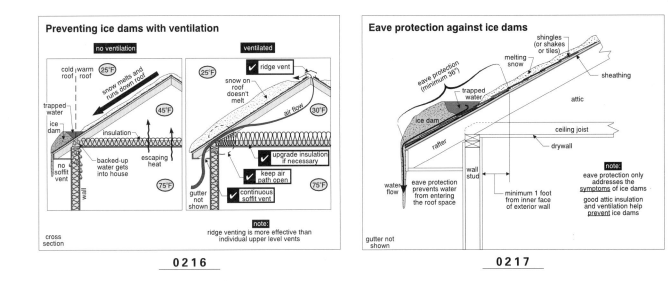

no ventilation

ventilated

cold roof / warm roof

25°F

snow melts and runs down roof

✓ ridge vent

25°F

snow on roof doesn't melt

trapped water

45°F

air flow

30°F

ice dam

insulation

no soffit vent

backed-up water gets into house

escaping heat

wall

75°F

✓ upgrade insulation if necessary

✓ keep air path open

✓ continuous soffit vent

gutter not shown

75°F

cross section

note:
ridge venting is more effective than individual upper level vents

0216

Eave protection against ice dams

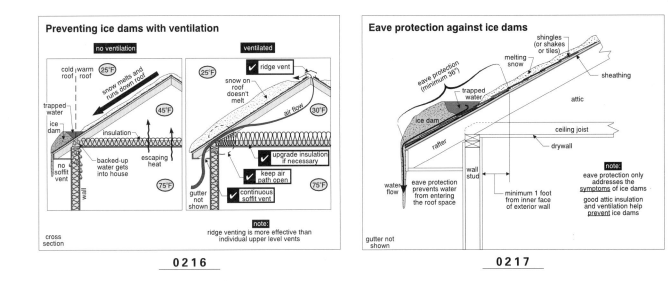

shingles (or shakes or tiles)

eave protection (minimum 36")

melting snow

sheathing

trapped water

attic

ice dam

rafter

ceiling joist

drywall

water flow

eave protection prevents water from entering the roof space

wall stud

minimum 1 foot from inner face of exterior wall

note:
eave protection only addresses the symptoms of ice dams

good attic insulation and ventilation help prevent ice dams

gutter not shown

0217

Avalanche guards

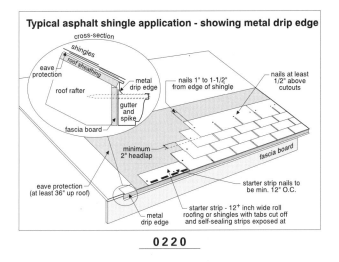

snow

avalanche guards (tab type)

avalanche guard (rail type)

slate roof

avalanche guards are more common on slate, clay and concrete roofs

0218

Asphalt shingle composition

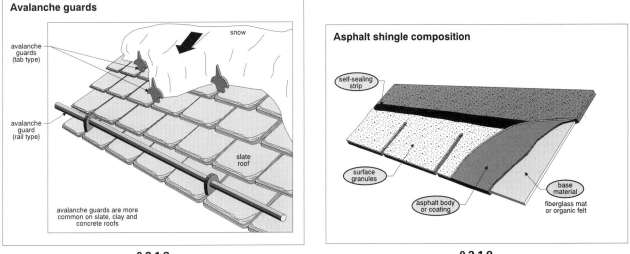

self-sealing strip

surface granules

asphalt body or coating

base material

fiberglass mat or organic felt

0219

Typical asphalt shingle application - showing metal drip edge

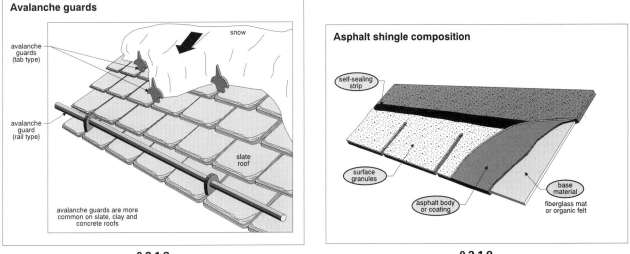

cross-section

shingles

roof sheathing

eave protection

metal drip edge

roof rafter

gutter and spike

fascia board

nails 1" to 1-1/2" from edge of shingle

nails at least 1/2" above cutouts

minimum 2" headlap

fascia board

eave protection (at least 36" up roof)

metal drip edge

starter strip nails to be min. 12" O.C.

starter strip - 12⁺ inch wide roll roofing or shingles with tabs cut off and self-sealing strips exposed at

0220

Self-sealing tabs

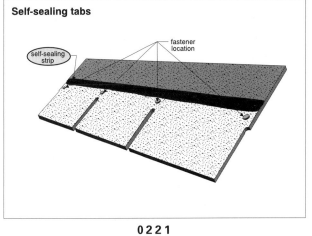

fastener location

self-sealing strip

0221

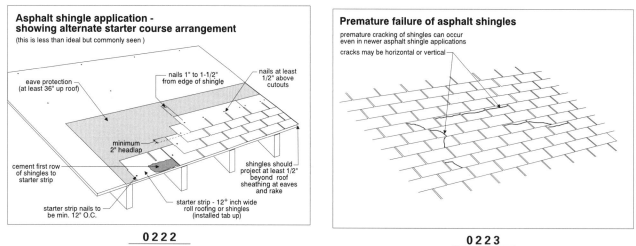

**Asphalt shingle application -
showing alternate starter course arrangement**

(this is less than ideal but commonly seen)

eave protection
(at least 36" up roof)

nails 1" to 1-1/2"
from edge of shingle

nails at least
1/2" above
cutouts

minimum
2" headlap

cement first row
of shingles to
starter strip

shingles should
project at least 1/2"
beyond roof
sheathing at eaves
and rake

starter strip nails to
be min. 12" O.C.

starter strip - 12⁺ inch wide
roll roofing or shingles
(installed tab up)

0222

Premature failure of asphalt shingles

premature cracking of shingles can occur
even in newer asphalt shingle applications

cracks may be horizontal or vertical

0223

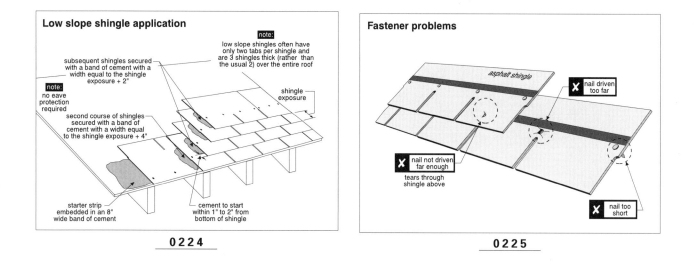

Low slope shingle application

note:
low slope shingles often have
only two tabs per shingle and
are 3 shingles thick (rather than
the usual 2) over the entire roof

subsequent shingles secured
with a band of cement with a
width equal to the shingle
exposure + 2"

shingle
exposure

note:
no eave
protection
required

second course of shingles
secured with a band of
cement with a width equal
to the shingle exposure + 4"

starter strip
embedded in an 8"
wide band of cement

cement to start
within 1" to 2" from
bottom of shingle

0224

Fastener problems

asphalt shingle

X nail driven
too far

X nail not driven
far enough

tears through
shingle above

X nail too
short

0225

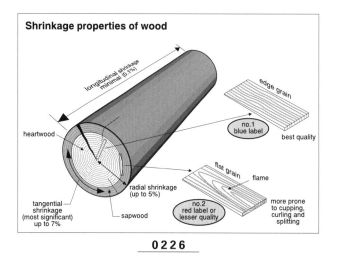

Shrinkage properties of wood

longitudinal shrinkage
minimal (0.1%)

edge grain

heartwood

no.1
blue label

best quality

flat grain

flame

radial shrinkage
(up to 5%)

no.2
red label or
lesser quality

more prone
to cupping,
curling and
splitting

tangential
shrinkage
(most significant)
up to 7%

sapwood

0226

Wood shingles and shakes

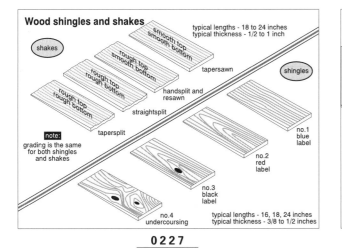

shakes

smooth top
smooth bottom

rough top
smooth bottom

rough top
rough bottom

rough top
rough bottom

tapersawn

handsplit and
resawn

straightsplit

tapersplit

typical lengths - 18 to 24 inches
typical thickness - 1/2 to 1 inch

shingles

note:
grading is the same
for both shingles
and shakes

no.1
blue
label

no.2
red
label

no.3
black
label

no.4
undercoursing

typical lengths - 16, 18, 24 inches
typical thickness - 3/8 to 1/2 inches

0227

Cedar shingle application
over plywood sheathing

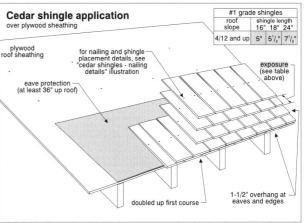

#1 grade shingles			
roof slope	shingle length		
	16"	18"	24"
4/12 and up	5"	5¹/₂"	7¹/₂"

plywood
roof sheathing

for nailing and shingle
placement details, see
"cedar shingles - nailing
details" illustration

eave protection
(at least 36" up roof)

exposure
(see table
above)

doubled up first course

1-1/2" overhang at
eaves and edges

0228

Cedar shingles - nailing details

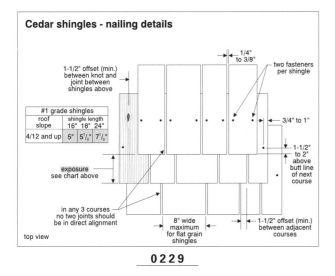

1-1/2" offset (min.)
between knot and
joint between
shingles above

1/4"
to 3/8"

two fasteners
per shingle

#1 grade shingles			
roof slope	shingle length		
	16"	18"	24"
4/12 and up	5"	5¹/₂"	7¹/₂"

3/4" to 1"

1-1/2"
to 2"
above
butt line
of next
course

exposure
see chart above

in any 3 courses
no two joints should
be in direct alignment

8" wide
maximum
for flat grain
shingles

1-1/2" offset (min.)
between adjacent
courses

top view

0229

Cedar shingle application-
spaced plank sheathing

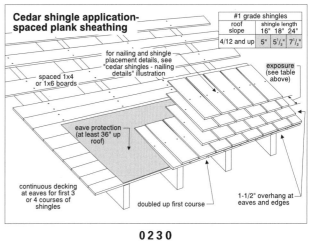

#1 grade shingles			
roof slope	shingle length		
	16"	18"	24"
4/12 and up	5"	5¹/₂"	7¹/₂"

for nailing and shingle
placement details, see
"cedar shingles - nailing
details" illustration

spaced 1x4
or 1x6 boards

exposure
(see table
above)

eave protection
(at least 36" up
roof)

continuous decking
at eaves for first 3
or 4 courses of
shingles

doubled up first course

1-1/2" overhang at
eaves and edges

0230

Cedar shake application

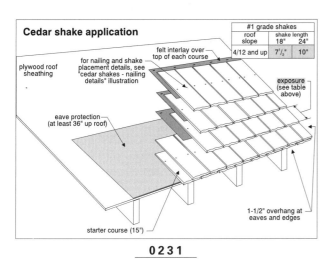

#1 grade shakes		
roof slope	shake length	
	18"	24"
4/12 and up	7¹/₂"	10"

felt interlay over
top of each course

plywood roof
sheathing

for nailing and shake
placement details, see
"cedar shakes - nailing
details" illustration

exposure
(see table
above)

eave protection
(at least 36" up roof)

starter course (15")

1-1/2" overhang at
eaves and edges

0231

Cedar shakes - nailing details

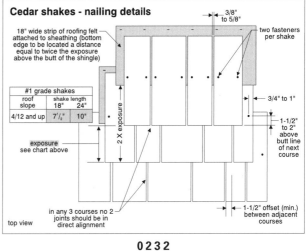

3/8"
to 5/8"

18" wide strip of roofing felt
attached to sheathing (bottom
edge to be located a distance
equal to twice the exposure
above the butt of the shingle)

two fasteners
per shake

#1 grade shakes		
roof slope	shake length	
	18"	24"
4/12 and up	7¹/₂"	10"

2 X exposure

3/4" to 1"

1-1/2"
to 2"
above
butt line
of next
course

exposure
see chart above

in any 3 courses no 2
joints should be in
direct alignment

1-1/2" offset (min.)
between adjacent
courses

top view

0232

Cedar shingle application- using synthetic ventilation layer

#1 grade shingles			
roof slope	shingle length		
	16"	18"	24"
4/12 and up	5"	5 1/8"	7 1/2"

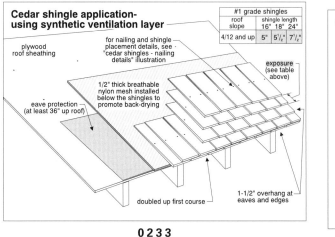

plywood roof sheathing

for nailing and shingle placement details, see "cedar shingles - nailing details" illustration

1/2" thick breathable nylon mesh installed below the shingles to promote back-drying

eave protection (at least 36" up roof)

exposure (see table above)

doubled up first course

1-1/2" overhang at eaves and edges

0233

Curling, cupping and splitting wood shingles

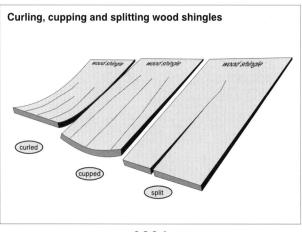

wood shingle   wood shingle   wood shingle

curled

cupped

split

0234

Slate types

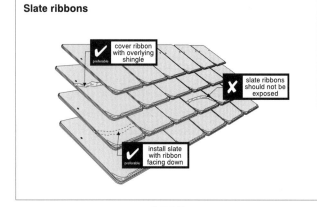

Pennsylvania slate

Vermont slate

Buckingham slate (Virginia)

typical life: 35 to 75 years

typical color: blue-grey, blue-black and black

typical life: 100 years

typical color: light-grey, grey-black, green, mottled purple and green, red (rare) and purple (rare)

typical life: 175 years

typical color: blue-grey to dark grey with an unusual luster

0235

Graduated slate roof

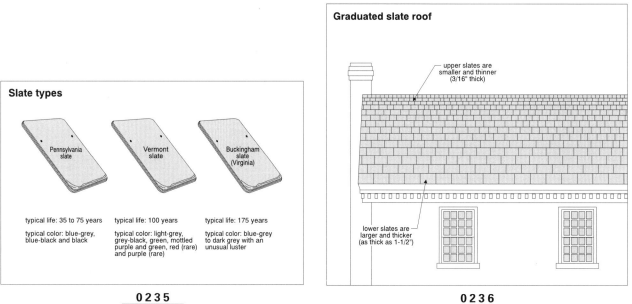

upper slates are smaller and thinner (3/16" thick)

lower slates are larger and thicker (as thick as 1-1/2")

0236

Slate ribbons

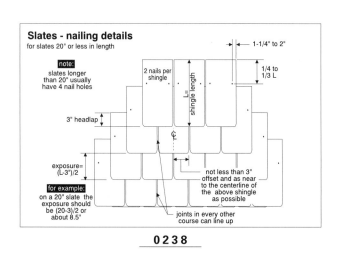

cover ribbon with overlying shingle
preferable

slate ribbons should not be exposed

install slate with ribbon facing down
preferable

0237

Slates - nailing details
for slates 20" or less in length

note: slates longer than 20" usually have 4 nail holes

2 nails per shingle

1-1/4" to 2"

1/4 to 1/3 L

shingle length L=

3" headlap

exposure= (L-3")/2

for example: on a 20" slate the exposure should be (20-3)/2 or about 8.5"

not less than 3" offset and as near to the centerline of the above shingle as possible

joints in every other course can line up

0238

Slate roof installation - dutch lap method

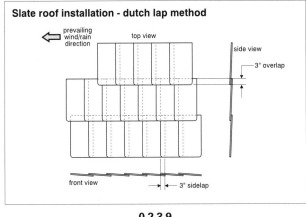

prevailing
wind/rain
direction

top view

side view

3" overlap

front view

3" sidelap

0239

Slate roof installation - french method

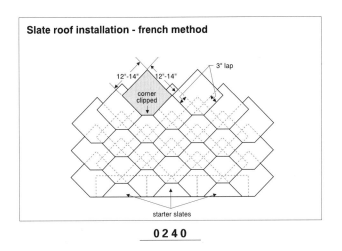

12"-14"   12"-14"

3" lap

corner
clipped

starter slates

0240

Slate repair methods- using a slate hook

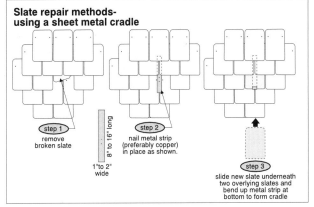

step 1

remove
broken slate

step 2

nail slate hook into
sheathing between two
slates so that hook is
level with bottom edge
of the new slate

step 3

slide new slate underneath
two overlying slates and
seat on slate hook

3"

stainless
steel
slate
hook

enlarged

0241

Slate repair methods- making babies

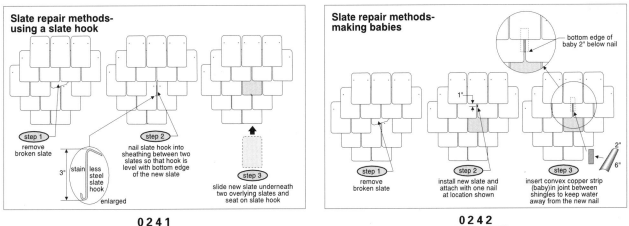

bottom edge of
baby 2" below nail

1"

step 1

remove
broken slate

step 2

install new slate and
attach with one nail
at location shown

step 3

insert convex copper strip
(baby)in joint between
shingles to keep water
away from the new nail

2"

6"

0242

Slate repair methods- using a sheet metal cradle

step 1

remove
broken slate

8" to 16" long

step 2

nail metal strip
(preferably copper)
in place as shown.

1"to 2"
wide

step 3

slide new slate underneath
two overlying slates and
bend up metal strip at
bottom to form cradle

0243

Clay tile - tapered mission style

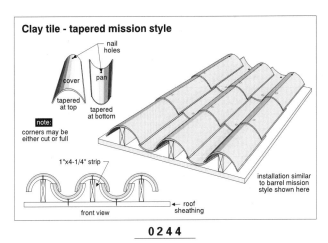

nail
holes

cover

pan

tapered
at top

tapered
at bottom

note:

corners may be
either cut or full

1"x4-1/4" strip

front view

roof
sheathing

installation similar
to barrel mission
style shown here

0244

Clay tile - barrel mission style (straight)

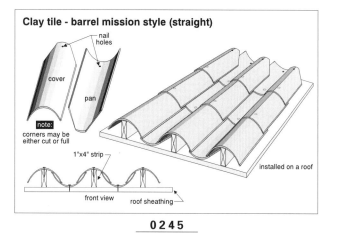

nail holes

cover

pan

note: corners may be either cut or full

1"x4" strip

front view

roof sheathing

installed on a roof

0245

Clay tile - "S" style

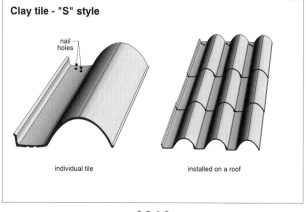

nail holes

individual tile

installed on a roof

0246

Clay tile - interlocking shingle

top view

side view

front view

installed on roof

front view

roof sheathing

0247

How clay tiles are secured

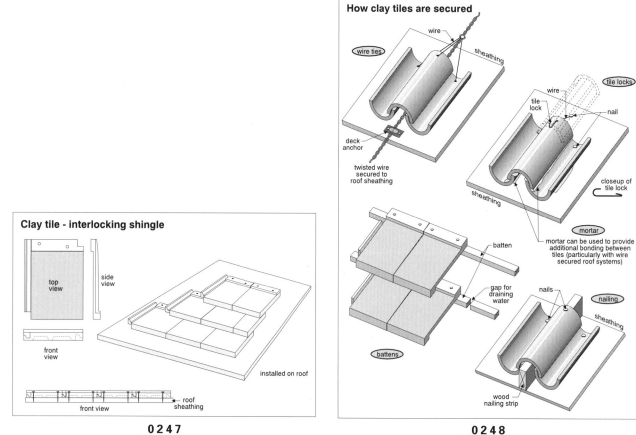

wire ties

wire

sheathing

deck anchor

twisted wire secured to roof sheathing

tile locks

wire

tile lock

nail

sheathing

closeup of tile lock

mortar

mortar can be used to provide additional bonding between tiles (particularly with wire secured roof systems)

batten

gap for draining water

nails

nailing

sheathing

battens

wood nailing strip

0248

Eave closures for spanish or mission tiles

spanish tile

roof sheathing

eave closure

bottom edge of roof

drain hole

fascia board

0249

Concrete tiles

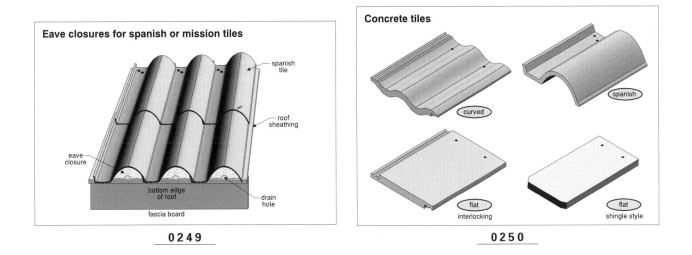

curved

spanish

flat
interlocking

flat
shingle style

0250

Cracked or broken tiles

the circled areas show where cracked or broken tiles are more likely to be found

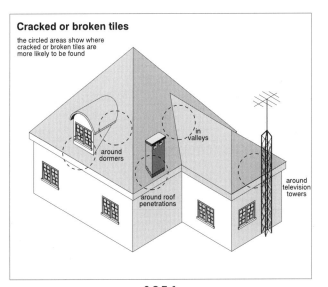

in valleys

around dormers

around roof penetrations

around television towers

0251

How concrete tiles are secured

battens

batten

gap for draining water

nailing

mortar

mortar may be used to secure tiles (typically in addition to nailing)

sheathing

hurricane clip

hurricane clips

0252

Fiber cement shingles

fiber cement shingles come in many shapes and colors

rectangular and diamond shapes are common

they are lighter and more brittle than clay or slate shingles

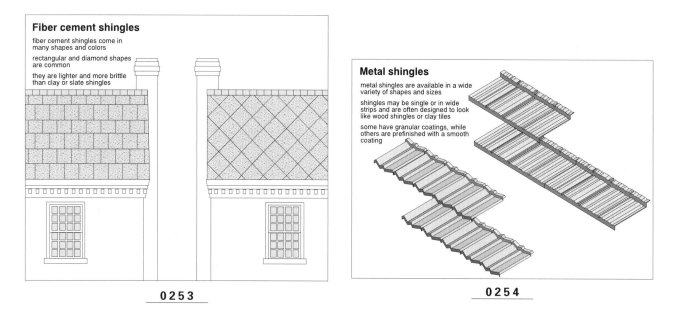

0253

Metal shingles

metal shingles are available in a wide variety of shapes and sizes

shingles may be single or in wide strips and are often designed to look like wood shingles or clay tiles

some have granular coatings, while others are prefinished with a smooth coating

0254

Sheet metal roofing

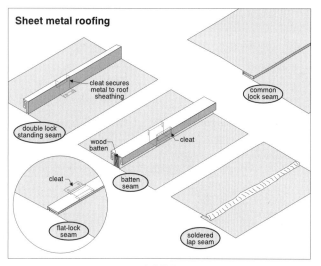

cleat secures metal to roof sheathing

common lock seam

double lock standing seam

wood batten

cleat

cleat

batten seam

flat-lock seam

soldered lap seam

0255

Roll roofing

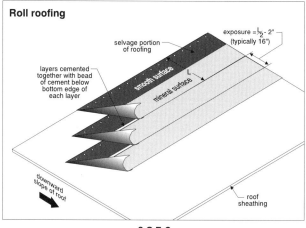

selvage portion of roofing

exposure = ½ - 2" (typically 16")

layers cemented together with bead of cement below bottom edge of each layer

smooth surface

mineral surface

downward slope of roof

roof sheathing

0256

Roll roofing problems

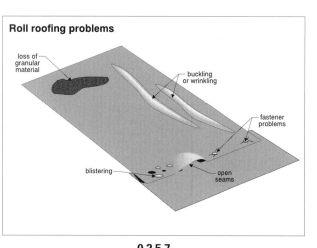

loss of granular material

buckling or wrinkling

fastener problems

blistering

open seams

0257

Roof valleys - open and closed

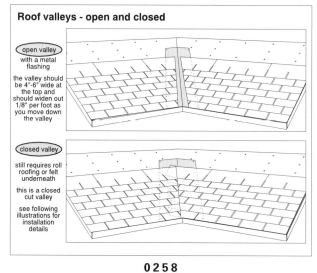

open valley
with a metal flashing

the valley should be 4"-6" wide at the top and should widen out 1/8" per foot as you move down the valley

closed valley
still requires roll roofing or felt underneath

this is a closed cut valley

see following illustrations for installation details

0258

Why valleys have a lower slope

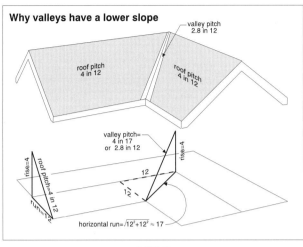

valley pitch 2.8 in 12

roof pitch 4 in 12

roof pitch 4 in 12

valley pitch= 4 in 17 or 2.8 in 12

rise=4

rise=4

roof pitch=4 in 12

run=12

12

12

horizontal run=$\sqrt{12^2+12^2} \approx 17$

0259

Valley flashing

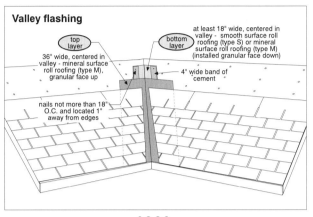

top layer

bottom layer

at least 18" wide, centered in valley - smooth surface roll roofing (type S) or mineral surface roll roofing (type M) (installed granular face down)

36" wide, centered in valley - mineral surface roll roofing (type M), granular face up

4" wide band of cement

nails not more than 18" O.C. and located 1" away from edges

0260

Valley flashing made of asphalt shingles

open valleys laid with overlapped individual shingles are likely to leak

this is poor practice

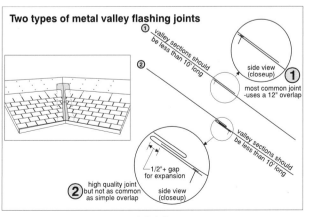

0261

Metal valley flashing

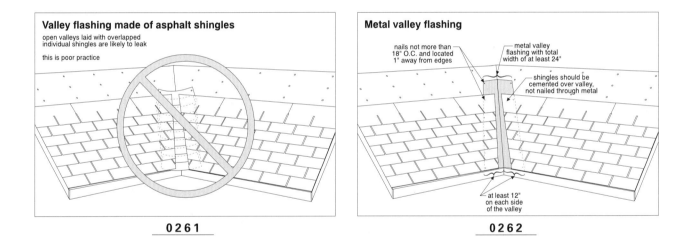

nails not more than 18" O.C. and located 1" away from edges

metal valley flashing with total width of at least 24"

shingles should be cemented over valley, not nailed through metal

at least 12" on each side of the valley

0262

Two types of metal valley flashing joints

① valley sections should be less than 10' long

② valley sections should be less than 10' long

side view (closeup) ①

most common joint -uses a 12" overlap

1/2"+ gap for expansion

② high quality joint but not as common as simple overlap

side view (closeup)

0263

Metal valley secured with cleats

this is a very good quality installation method but is not commonly seen

cleats spaced every 12" O.C.

1/2" return

metal valley flashing with total width of at least 24"

shingles should be cemented over valley, not nailed through metal

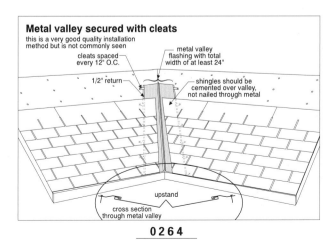

upstand

cross section through metal valley

0264

Metal valley flashing - alternate application

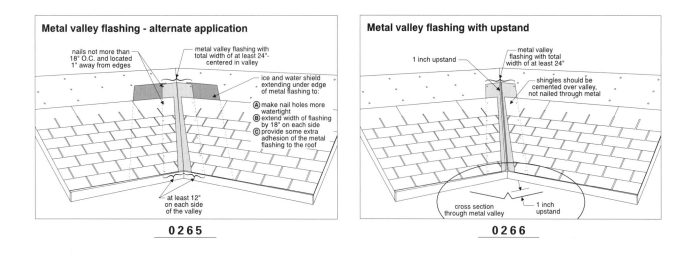

nails not more than 18" O.C. and located 1" away from edges

metal valley flashing with total width of at least 24"- centered in valley

ice and water shield extending under edge of metal flashing to:
Ⓐ make nail holes more watertight
Ⓑ extend width of flashing by 18" on each side
Ⓒ provide some extra adhesion of the metal flashing to the roof

at least 12" on each side of the valley

0265

Metal valley flashing with upstand

1 inch upstand

metal valley flashing with total width of at least 24"

shingles should be cemented over valley, not nailed through metal

cross section through metal valley

1 inch upstand

0266

Valley flashings - cutting the points

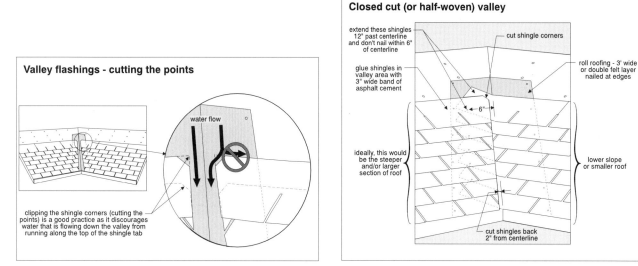

water flow

clipping the shingle corners (cutting the points) is a good practice as it discourages water that is flowing down the valley from running along the top of the shingle tab

0267

Closed cut (or half-woven) valley

extend these shingles 12" past centerline and don't nail within 6" of centerline

cut shingle corners

glue shingles in valley area with 3" wide band of asphalt cement

roll roofing - 3' wide or double felt layer nailed at edges

6"

ideally, this would be the steeper and/or larger section of roof

lower slope or smaller roof

cut shingles back 2" from centerline

0268

Fully woven closed valley

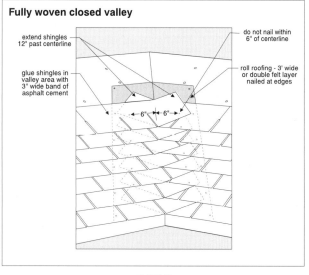

extend shingles 12" past centerline

do not nail within 6" of centerline

glue shingles in valley area with 3" wide band of asphalt cement

roll roofing - 3' wide or double felt layer nailed at edges

6"   6"

0269

Chimney flashings - overview

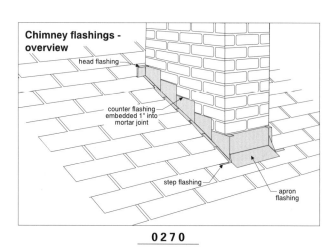

head flashing

counter flashing embedded 1" into mortar joint

step flashing

apron flashing

0270

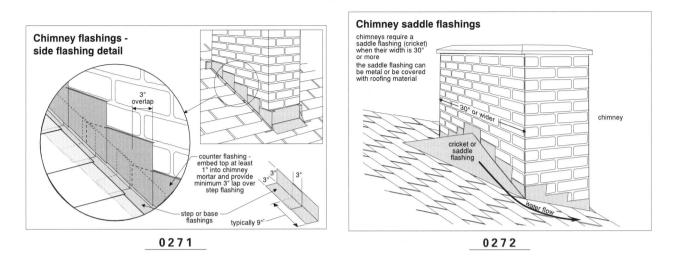

Chimney flashings - side flashing detail

3" overlap

counter flashing - embed top at least 1" into chimney mortar and provide minimum 3" lap over step flashing

step or base flashings

3" 3" 3"

typically 9"

0271

Chimney saddle flashings

chimneys require a saddle flashing (cricket) when their width is 30" or more
the saddle flashing can be metal or be covered with roofing material

30" or wider

cricket or saddle flashing

chimney

water flow

0272

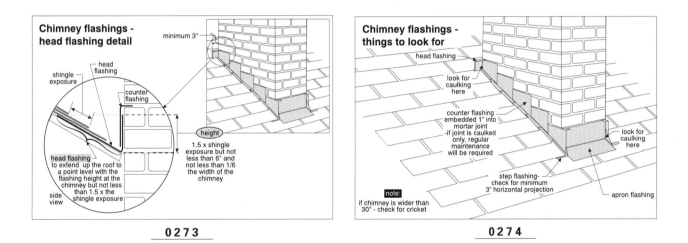

Chimney flashings - head flashing detail

minimum 3"

shingle exposure

head flashing

counter flashing

head flashing to extend up the roof to a point level with the flashing height at the chimney but not less than 1.5 x shingle exposure

side view

height

1.5 x shingle exposure but not less than 6" and not less than 1/6 the width of the chimney

0273

Chimney flashings - things to look for

head flashing

look for caulking here

counter flashing embedded 1" into mortar joint -if joint is caulked only, regular maintenance will be required

look for caulking here

step flashing- check for minimum 3" horizontal projection

apron flashing

note:
if chimney is wider than 30" - check for cricket

0274

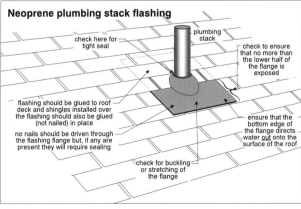

Neoprene plumbing stack flashing

check here for tight seal

plumbing stack

check to ensure that no more than the lower half of the flange is exposed

flashing should be glued to roof deck and shingles installed over the flashing should also be glued (not nailed) in place

no nails should be driven through the flashing flange but, if any are present they will require sealing

ensure that the bottom edge of the flange directs water out onto the surface of the roof

check for buckling or stretching of the flange

0275

Plumbing stack flashing- stretched or buckled

plumbing stack

stretched
roof drops down

plumbing stack

potential water entry point

buckled
stack drops down

potential water entry points

cross section

0276

Roof/sidewall flashings

sidings such as wood, metal, vinyl or stucco can serve as counter flashing

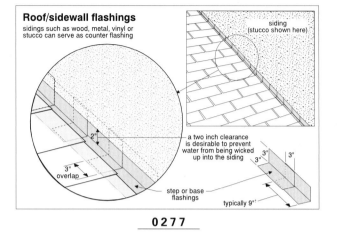

siding (stucco shown here)

a two inch clearance is desirable to prevent water from being wicked up into the siding

2"

3" overlap

step or base flashings

typically 9"

3"   3"   3"

0277

Roof/sidewall flashings - clay tile

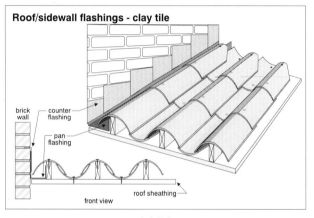

brick wall

counter flashing

pan flashing

roof sheathing

front view

0278

Roof/masonry sidewall flashings

base flashings can be nailed to roof (preferable) or wall to allow for differential movement but, they should not be nailed to both

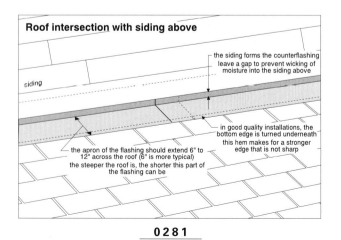

3" overlap

3" overlap

counter flashing or side cap flashing - embed top at least 1" into chimney mortar and lap bottom over step flashing

step or base flashings

typically 9"

3"   3"   3"

0279

Roof intersection with brick wall above

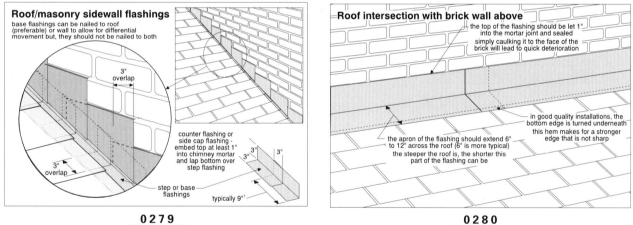

the top of the flashing should be let 1" into the mortar joint and sealed

simply caulking it to the face of the brick will lead to quick deterioration

in good quality installations, the bottom edge is turned underneath this hem makes for a stronger edge that is not sharp

the apron of the flashing should extend 6" to 12" across the roof (6" is more typical) the steeper the roof is, the shorter this part of the flashing can be

0280

Roof intersection with siding above

siding

the siding forms the counterflashing leave a gap to prevent wicking of moisture into the siding above

in good quality installations, the bottom edge is turned underneath this hem makes for a stronger edge that is not sharp

the apron of the flashing should extend 6" to 12" across the roof (6" is more typical) the steeper the roof is, the shorter this part of the flashing can be

0281

Ridge shingle application

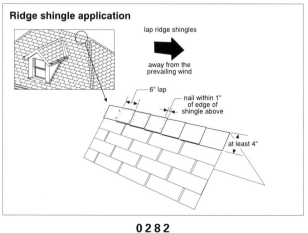

lap ridge shingles

away from the prevailing wind

6" lap

nail within 1" of edge of shingle above

at least 4"

0282

Skylight in sloped roof - curb mount

skylight
roof rafter
skylight well
ceiling joist

skylight frame
head flashing
shingle
header
curb - 4" min.
channel for collecting condensation
apron flashing
exterior wall

cross section

0283

Skylight in sloped roof - curbless

curbless skylights are low quality units prone to leakage
skylight
roof rafter
skylight well
ceiling joist

shingle
header
skylight
exterior wall

cross section

0284

Skylight in sloped roof - integral curb (self-flashing skylight)

the integral flashings on these skylights are small making the units more prone to leakage
skylight
roof rafter
skylight well
ceiling joist

skylight frame
shingle
header
integral curb/flashings
channel for collecting condensation
exterior wall

cross section

0285

Ice damming at skylight

snow
skylight
water buildup
ice dam
water entry
heat loss
roof rafter
snow
ceiling joist
exterior wall

localized heat loss causes snow to melt around the skylight
upon hitting the colder roof below the skylight, the water refreezes - building up a dam
water susequently running down the roof can back up under the shingles or skylight

cross section

0286

Solariums - areas to watch for

pay particular attention to intersections of the solarium roof with the walls of the house

joints around the glass are a second area of vulnerability
check the condition of the sealant (new caulking may

0287

Dormer siding flashings

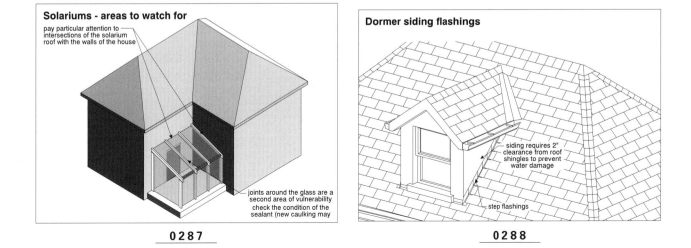

siding requires 2" clearance from roof shingles to prevent water damage
step flashings

0288

Flat roof drainage systems

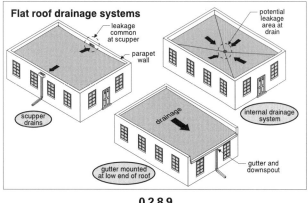

leakage common at scupper

parapet wall

potential leakage area at drain

scupper drains

internal drainage system

drainage

gutter mounted at low end of roof

gutter and downspout

0289

Inverted Roof Membrane Assembly (IRMA)

also called protected membrane

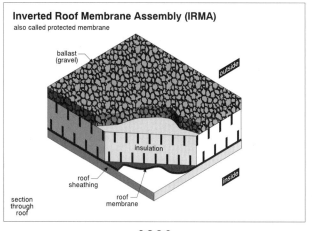

ballast (gravel)

outside

insulation

roof sheathing

roof membrane

inside

section through roof

0290

Built up roofing membrane - 4 ply

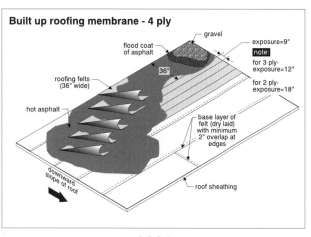

flood coat of asphalt

gravel

exposure=9"

note:

for 3 ply- exposure=12"

for 2 ply- exposure=18"

36"

roofing felts (36" wide)

hot asphalt

base layer of felt (dry laid) with minimum 2" overlap at edges

downward slope of roof

roof sheathing

0291

Damaged and patched flat roofs

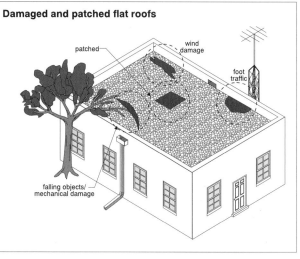

patched

wind damage

foot traffic

falling objects/ mechanical damage

0292

Blisters in built up roofing membranes

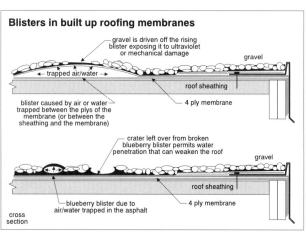

gravel is driven off the rising blister exposing it to ultraviolet or mechanical damage

gravel

trapped air/water

roof sheathing

4 ply membrane

blister caused by air or water trapped between the plys of the membrane (or between the sheathing and the membrane)

crater left over from broken blueberry blister permits water penetration that can weaken the roof

gravel

roof sheathing

4 ply membrane

blueberry blister due to air/water trapped in the asphalt

cross section

0293

Flat roof alligatoring

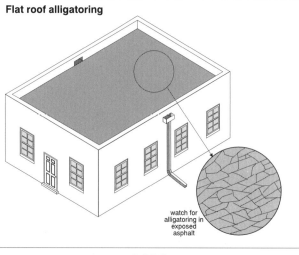

watch for alligatoring in exposed asphalt

0294

Gravel erosion (scouring) on flat roofs

gravel scouring can occur at the windward corner of roofs or where downspouts discharge onto flat roofs

loss of gravel can lead to quick deterioration of the roof membrane

wind

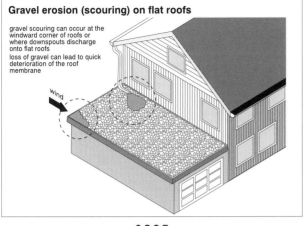

0295

Ridging and fishmouths on flat roofs

roofing felt

ridges and wrinkles may be due to excess asphalt moisture, felt slippage or differential thermal expansion

fishmouths are open at the edge of the felt ply potentially allowing direct water access

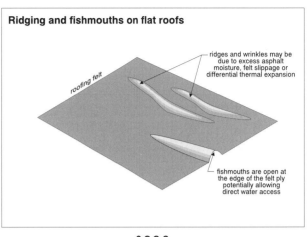

0296

Ponding on flat roofs

any roof that still has water on it after 48 hours is defined as a ponding roof

a dry roof may show signs of ponding - dirty circles on the roof or algae/vegetation growth

ponding water

scupper drain

sagging roof joists can render perimeter drains ineffective by allowing water to pond in the middle of the roof

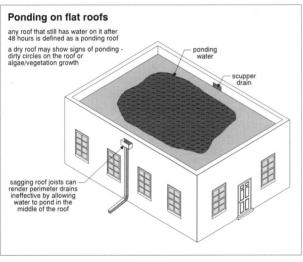

0297

Single ply modified bitumen roof

surface of membrane should be protected from ultraviolet light by:
-a granular surface
-liquid coating (eg. reflective)
-a foil surface
-ballast (eg. gravel)

36"

seams may be mopped or torched

downward slope of roof

roof sheathing

typically installed with a 3" overlap at the edges and sides

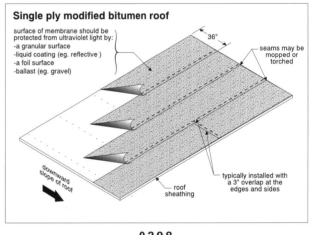

0298

Two-ply modified bitumen roof

cap sheet
-may have a granular surface on top
-typically laid with a 3" overlap at edges and sides

36"

seams may be mopped or torched

downward slope of roof

roof sheathing

base sheet (has no top coating) installed with a 3" overlap at the edges and sides

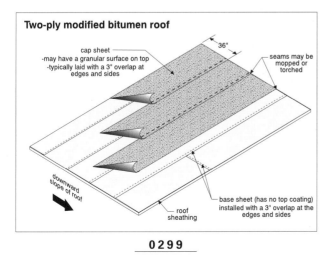

0299

Staggering side laps in single ply modified bitumen roofs

the side laps of adjacent rows should be staggered so that stresses do not concentrate along a single line
watch for opened seams if the laps have not been staggered

36"

seams may be mopped or torched

downward slope of roof

roof sheathing

typically installed with a 3" overlap at the edges and sides

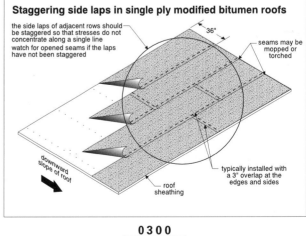

0300

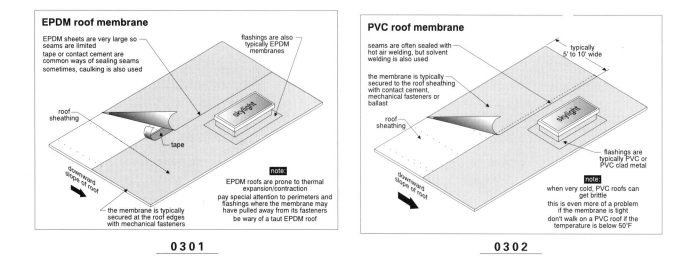

EPDM roof membrane

EPDM sheets are very large so seams are limited

tape or contact cement are common ways of sealing seams

sometimes, caulking is also used

flashings are also typically EPDM membranes

roof sheathing

skylight

tape

downward slope of roof

the membrane is typically secured at the roof edges with mechanical fasteners

note:
EPDM roofs are prone to thermal expansion/contraction

pay special attention to perimeters and flashings where the membrane may have pulled away from its fasteners

be wary of a taut EPDM roof

0301

PVC roof membrane

seams are often sealed with hot air welding, but solvent welding is also used

the membrane is typically secured to the roof sheathing with contact cement, mechanical fasteners or ballast

roof sheathing

typically 5' to 10' wide

skylight

flashings are typically PVC or PVC clad metal

note:
when very cold, PVC roofs can get brittle

this is even more of a problem if the membrane is tight

don't walk on a PVC roof if the temperature is below 50°F

downward slope of roof

0302

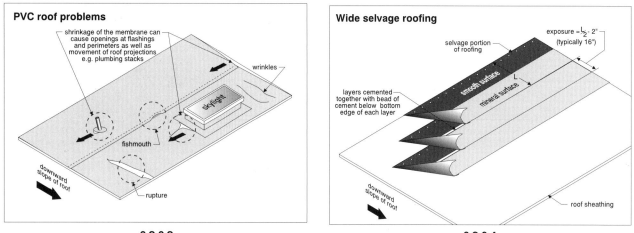

PVC roof problems

shrinkage of the membrane can cause openings at flashings and perimeters as well as movement of roof projections e.g. plumbing stacks

wrinkles

skylight

fishmouth

downward slope of roof

rupture

0303

Wide selvage roofing

selvage portion of roofing

exposure = $\frac{1}{2}$ - 2" (typically 16")

smooth surface

mineral surface

layers cemented together with bead of cement below bottom edge of each layer

downward slope of roof

roof sheathing

0304

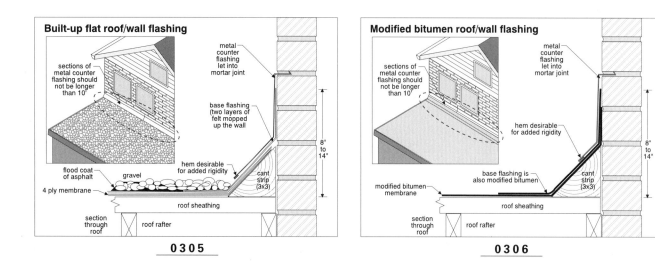

Built-up flat roof/wall flashing

sections of metal counter flashing should not be longer than 10'

metal counter flashing let into mortar joint

base flashing (two layers of felt mopped up the wall)

hem desirable for added rigidity

flood coat of asphalt

gravel

4 ply membrane

cant strip (3x3)

roof sheathing

section through roof

roof rafter

8" to 14"

0305

Modified bitumen roof/wall flashing

sections of metal counter flashing should not be longer than 10'

metal counter flashing let into mortar joint

hem desirable for added rigidity

modified bitumen membrane

base flashing is also modified bitumen

cant strip (3x3)

roof sheathing

section through roof

roof rafter

8" to 14"

0306

Two-piece counter flashing

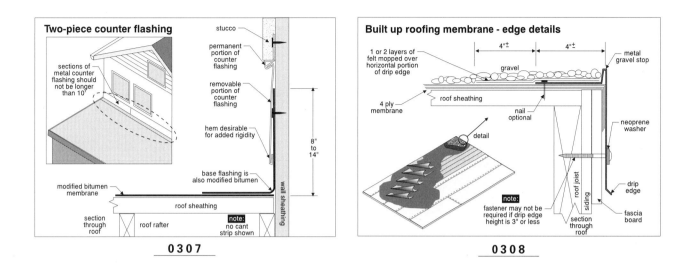

stucco

permanent portion of counter flashing

sections of metal counter flashing should not be longer than 10'

removable portion of counter flashing

hem desirable for added rigidity

8" to 14"

base flashing is also modified bitumen

modified bitumen membrane

roof sheathing

wall sheathing

section through roof

roof rafter

note: no cant strip shown

0307

Built up roofing membrane - edge details

1 or 2 layers of felt mopped over horizontal portion of drip edge

4"±   4"±

gravel

metal gravel stop

roof sheathing

4 ply membrane

nail optional

detail

neoprene washer

note: fastener may not be required if drip edge height is 3" or less

roof joist

siding

drip edge

fascia board

section through roof

0308

Built up roofing membrane - curbed edge

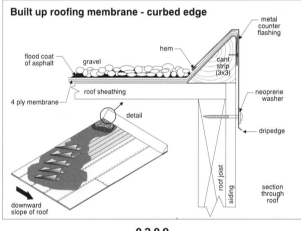

metal counter flashing

hem

flood coat of asphalt

gravel

cant strip (3x3)

roof sheathing

4 ply membrane

neoprene washer

detail

dripedge

roof joist

siding

downward slope of roof

section through roof

0309

Alternative attachment method for counter flashing

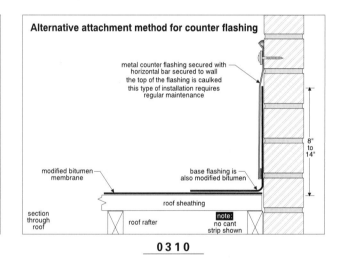

metal counter flashing secured with horizontal bar secured to wall the top of the flashing is caulked this type of installation requires regular maintenance

8" to 14"

modified bitumen membrane

base flashing is also modified bitumen

roof sheathing

section through roof

roof rafter

note: no cant strip shown

0310

Lower quality built-up flat roof/wall flashing

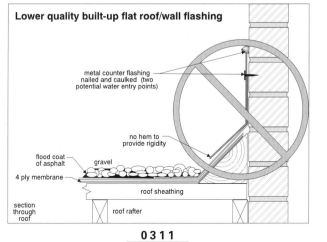

metal counter flashing nailed and caulked (two potential water entry points)

no hem to provide rigidity

flood coat of asphalt

gravel

4 ply membrane

roof sheathing

section through roof

roof rafter

0311

EPDM (or PVC) roof/wall flashing

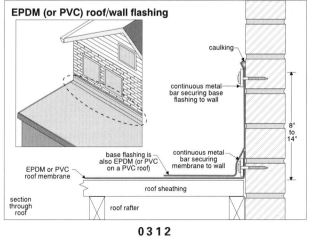

caulking

continuous metal bar securing base flashing to wall

8" to 14"

base flashing is also EPDM (or PVC on a PVC roof)

continuous metal bar securing membrane to wall

EPDM or PVC roof membrane

roof sheathing

section through roof

roof rafter

0312

Through-wall flashings

the through-wall flashing is designed to prevent water that enters the top of the wall from finding its way down into the building

this is a high quality detail not often found residentially

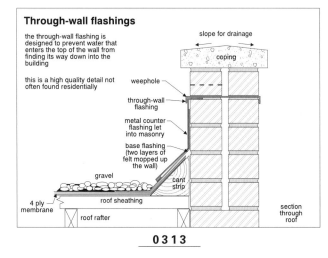

coping

slope for drainage

weephole

through-wall flashing

metal counter flashing let into masonry

base flashing (two layers of felt mopped up the wall)

cant strip

gravel

4 ply membrane

roof sheathing

roof rafter

section through roof

0 3 1 3

Metal clad parapet wall

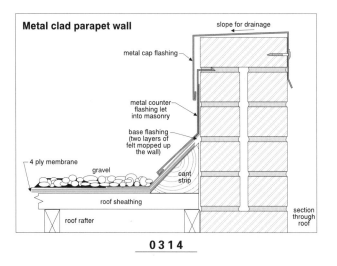

slope for drainage

metal cap flashing

metal counter flashing let into masonry

base flashing (two layers of felt mopped up the wall)

4 ply membrane

gravel

cant strip

roof sheathing

roof rafter

section through roof

0 3 1 4

Chimney cap details

seal any cracks or gaps in the cap

drainage

chimney

look for caulking here

cap

ideally, a groove is provided here to prevent water from running off the cap and down the chimney face

clay tile flue liner

cap should project at least 1" beyond chimney face

cross section

0 3 1 5

Skylight in flat roof

skylight

roof rafter

skylight frame

caulk/seal around head of fastener

counter flashing

base flashing

channel for collecting condensation

8" curb

cant strip

exterior wall

0 3 1 6

"Stack jack" plumbing stack flashing

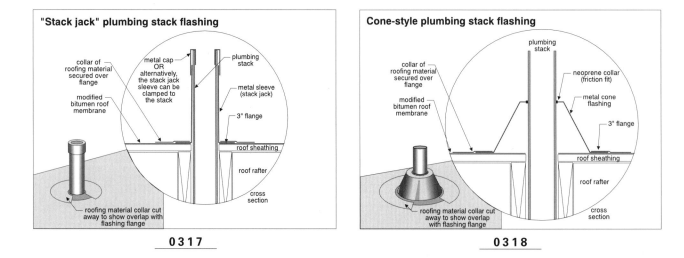

collar of roofing material secured over flange

metal cap OR alternatively, the stack jack sleeve can be clamped to the stack

plumbing stack

metal sleeve (stack jack)

modified bitumen roof membrane

3" flange

roof sheathing

roof rafter

cross section

roofing material collar cut away to show overlap with flashing flange

0 3 1 7

Cone-style plumbing stack flashing

plumbing stack

collar of roofing material secured over flange

neoprene collar (friction fit)

metal cone flashing

modified bitumen roof membrane

3" flange

roof sheathing

roof rafter

cross section

roofing material collar cut away to show overlap with flashing flange

0 3 1 8

Pitch pans for irregular roof penetrations

pitch pans or pockets are a less than ideal approach to sealing around irregular roof penetrations but, there are few alternatives

regular maintenance is required

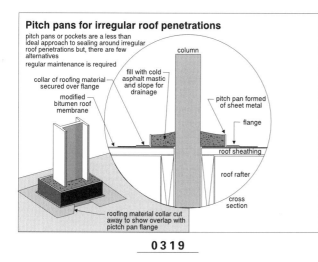

collar of roofing material secured over flange

modified bitumen roof membrane

column

fill with cold asphalt mastic and slope for drainage

pitch pan formed of sheet metal

flange

roof sheathing

roof rafter

cross section

roofing material collar cut away to show overlap with pitch pan flange

0 3 1 9

Drain in modified bitumen roof

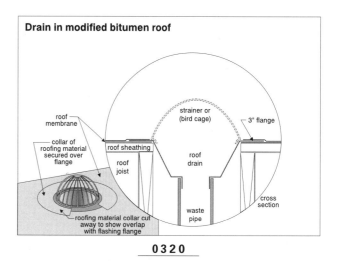

roof membrane

collar of roofing material secured over flange

strainer or (bird cage)

3" flange

roof sheathing

roof joist

roof drain

waste pipe

cross section

roofing material collar cut away to show overlap with flashing flange

0 3 2 0

Flat roof draining onto sloped roof

4"±   4"±

gravel

roof sheathing

metal gravel stop

hem for added rigidity

shingles

1 or 2 layers of felt mopped over horizontal portion of drip edge

4 ply membrane

roof rafter

roof sheathing

roof rafter

section through roof

0 3 2 1

Sloped roof draining onto flat roof

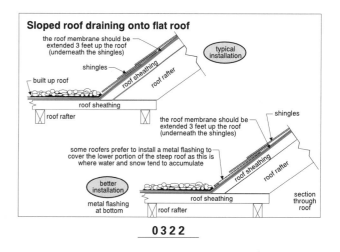

the roof membrane should be extended 3 feet up the roof (underneath the shingles)

shingles

built up roof

roof sheathing

roof rafter

typical installation

roof sheathing

roof rafter

the roof membrane should be extended 3 feet up the roof (underneath the shingles)

shingles

some roofers prefer to install a metal flashing to cover the lower portion of the steep roof as this is where water and snow tend to accumulate

better installation

metal flashing at bottom

roof rafter

roof sheathing

roof rafter

section through roof

0 3 2 2

PART 3

THE EXTERIOR

ARCHITECTURAL STYLES

BUILDING SHAPES AND DETAILS

WINDOWS

COLUMNS

SPECIFIC HOUSE STYLES

EXTERIOR CLADDING

GENERAL

WALL SURFACES - GENERAL

MASONRY

STUCCO

Linear plan vs. massed plan

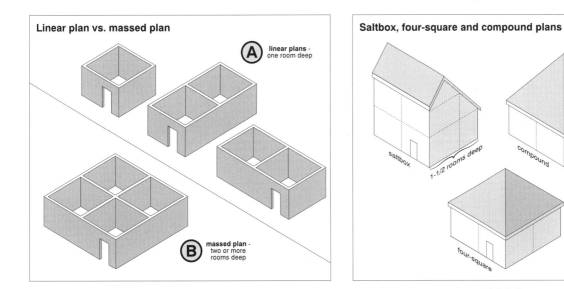

(A) **linear plans -**
one room deep

(B) **massed plan -**
two or more
rooms deep

0323

Saltbox, four-square and compound plans

saltbox

1-1/2 rooms deep

compound

four-square

0324

Flat roofs

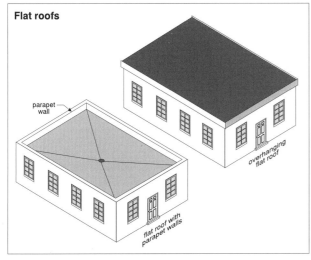

parapet
wall

overhanging
flat roof

flat roof with
parapet walls

0325

Bell-cast eave

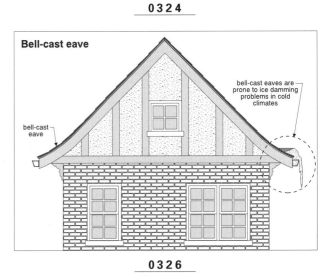

bell-cast eaves are
prone to ice damming
problems in cold
climates

bell-cast
eave

0326

Roof details

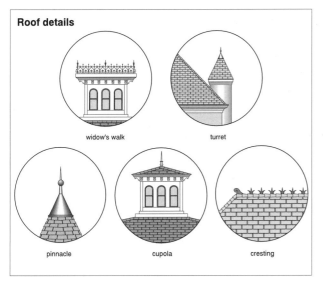

widow's walk

turret

pinnacle

cupola

cresting

0327

Gable details

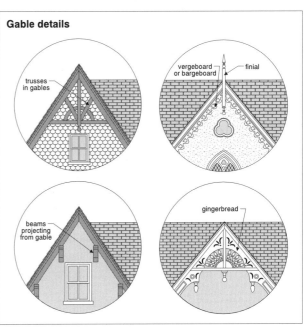

trusses
in gables

vergeboard
or bargeboard

finial

beams
projecting
from gable

gingerbread

0328

Decorative chimneys

decorative chimneys are
often found in Tudor and
Queen Anne houses

chimney
pot

decorative
masonry

0329

Cornice terms

cornice

frieze } entablature

architrave

these are formal definitions

many people use the word **cornice**
to refer to the whole decorated
area between the wall and the roof

0330

Dormer types

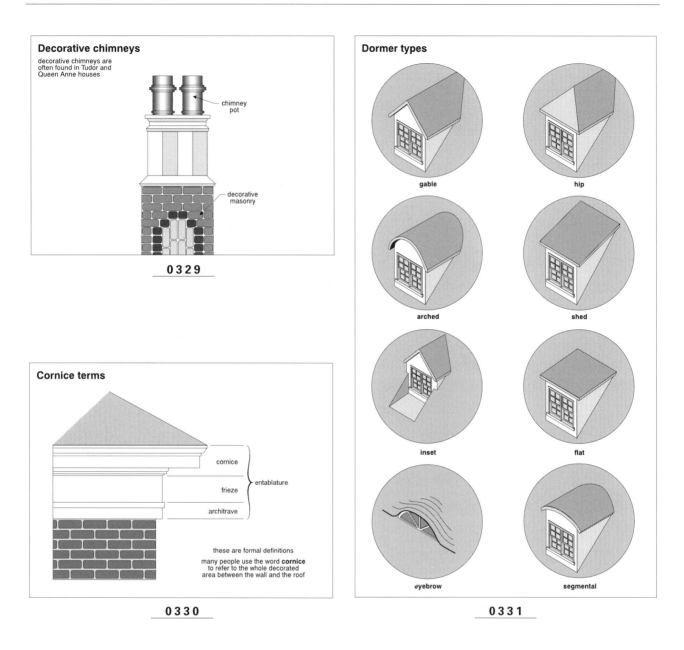

gable

hip

arched

shed

inset

flat

eyebrow

segmental

0331

Brackets versus dentils

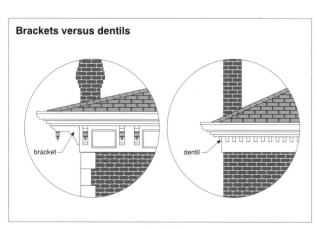

bracket

dentil

0332

Half-timbering

half-timbering and
nogging are commonly
found on Tudor homes

half-timbering

nogging

0333

Quoining

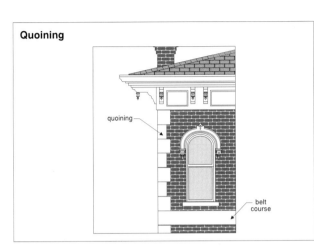

quoining

belt course

0 3 3 4

Double and single-hung windows

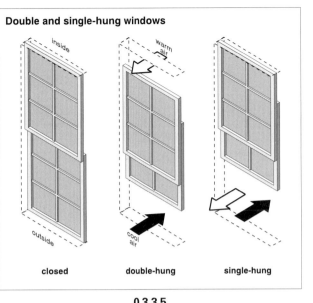

inside

warm air

outside

cool air

closed   **double-hung**   **single-hung**

0 3 3 5

Casement windows

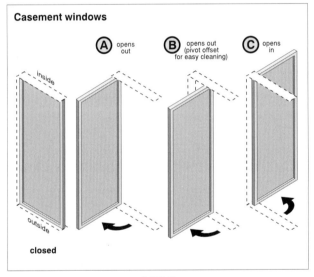

Ⓐ opens out   Ⓑ opens out (pivot offset for easy cleaning)   Ⓒ opens in

inside

outside

closed

0 3 3 6

Sliders

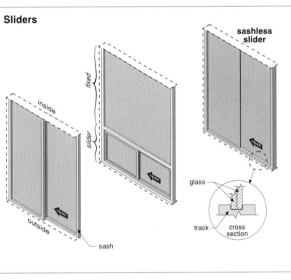

sashless slider

inside

fixed

slider

outside

sash

glass

track   cross section

0 3 3 7

Awning and hopper windows

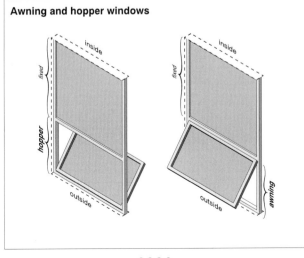

inside

fixed

hopper

outside

inside

fixed

awning

outside

0 3 3 8

Jalousie (or louver) windows

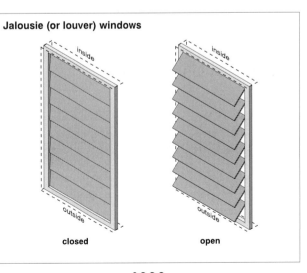

inside

inside

outside

outside

closed   **open**

0 3 3 9

Window terms

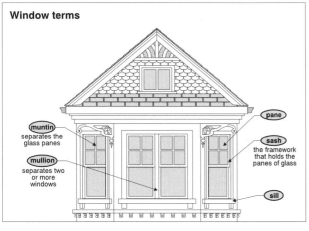

muntin
separates the glass panes

mullion
separates two or more windows

pane

sash
the framework that holds the panes of glass

sill

0 3 4 0

Window shapes

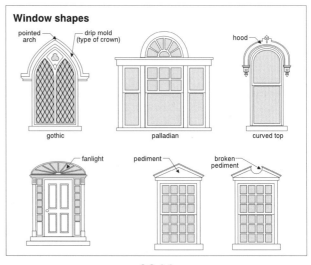

pointed arch

drip mold (type of crown)

gothic

palladian

hood

curved top

fanlight

pediment

broken pediment

0 3 4 1

Bow, bay and oriel windows

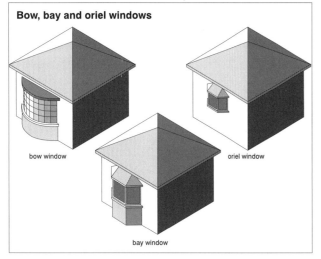

bow window

oriel window

bay window

0 3 4 2

Door lights

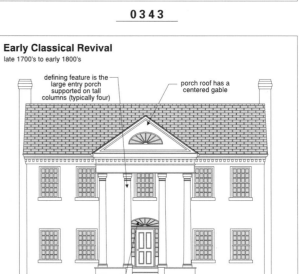

fanlight

pediment

transom light

side light

pilaster

0 3 4 3

Columns

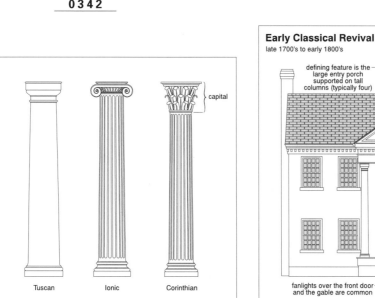

capital

Doric

Tuscan

Ionic

Corinthian

0 3 4 4

Early Classical Revival
late 1700's to early 1800's

defining feature is the large entry porch supported on tall columns (typically four)

porch roof has a centered gable

fanlights over the front door and the gable are common

front facade is symmetrical with windows lining up horizontally and vertically

0 3 4 5

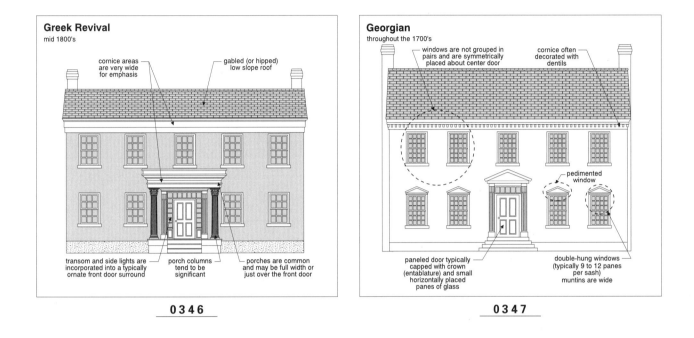

Greek Revival
mid 1800's

cornice areas are very wide for emphasis

gabled (or hipped) low slope roof

transom and side lights are incorporated into a typically ornate front door surround

porch columns tend to be significant

porches are common and may be full width or just over the front door

0346

Georgian
throughout the 1700's

windows are not grouped in pairs and are symmetrically placed about center door

cornice often decorated with dentils

pedimented window

paneled door typically capped with crown (entablature) and small horizontally placed panes of glass

double-hung windows (typically 9 to 12 panes per sash)
muntins are wide

0347

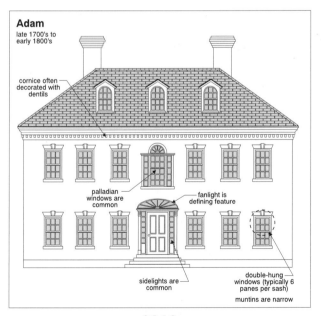

Adam
late 1700's to early 1800's

cornice often decorated with dentils

palladian windows are common

fanlight is defining feature

sidelights are common

double-hung windows (typically 6 panes per sash)
muntins are narrow

0348

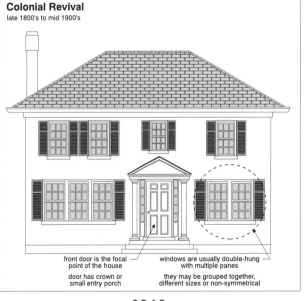

Colonial Revival
late 1800's to mid 1900's

front door is the focal point of the house

door has crown or small entry porch

windows are usually double-hung with multiple panes

they may be grouped together, different sizes or non-symmetrical

0349

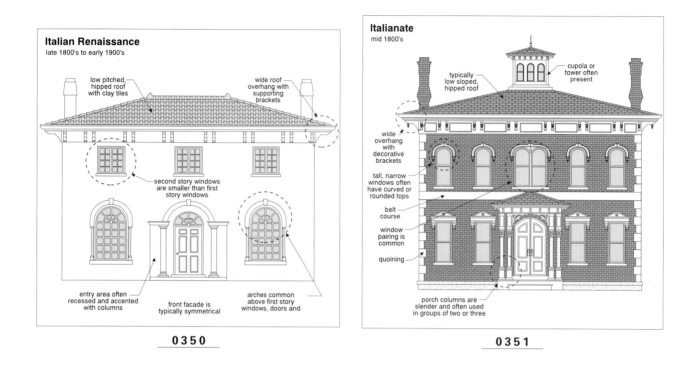

Italian Renaissance
late 1800's to early 1900's

low pitched, hipped roof with clay tiles

wide roof overhang with supporting brackets

second story windows are smaller than first story windows

entry area often recessed and accented with columns

front facade is typically symmetrical

arches common above first story windows, doors and

0 3 5 0

Italianate
mid 1800's

cupola or tower often present

typically low sloped, hipped roof

wide overhang with decorative brackets

tall, narrow windows often have curved or rounded tops

belt course

window pairing is common

quoining

porch columns are slender and often used in groups of two or three

0 3 5 1

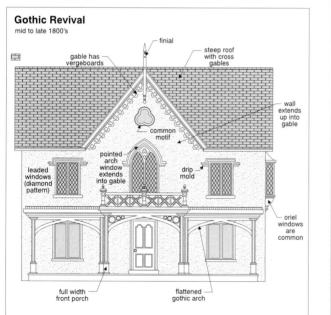

Gothic Revival
mid to late 1800's

finial

gable has vergeboards

steep roof with cross gables

wall extends up into gable

common motif

pointed arch window extends into gable

leaded windows (diamond pattern)

drip mold

oriel windows are common

full width front porch

flattened gothic arch

0 3 5 2

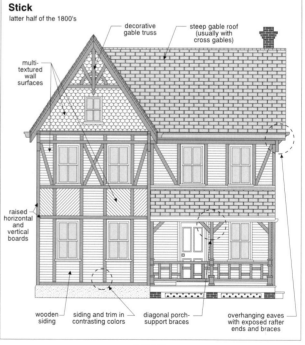

Stick
latter half of the 1800's

decorative gable truss

steep gable roof (usually with cross gables)

multi-textured wall surfaces

raised horizontal and vertical boards

wooden siding

siding and trim in contrasting colors

diagonal porch-support braces

overhanging eaves with exposed rafter ends and braces

0 3 5 3

Queen Anne
late 1800's to early 1900's

complex, steep roof with large front facing gable

roof cresting (and finials) are common

roof is often hipped

towers are an identifying feature

lines of windows

smaller single pane above larger one

front facade is asymmetrical

porches typically wrap around two sides of the house

spindlework and intricate columns are very common

0354

Additional Queen Anne details
late 1800's to early 1900's

large, patterned chimney

gable ornament

patterned shingle work

pent roof

corner bracket

brackets or other devices to avoid smooth-looking walls

large pane of glass surrounded by smaller panes

0355

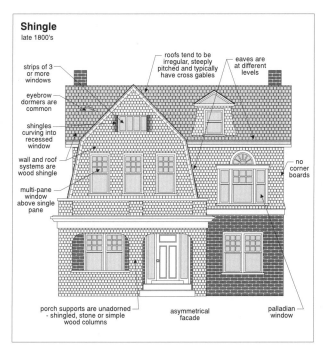

Shingle
late 1800's

strips of 3 or more windows

eyebrow dormers are common

shingles curving into recessed window

wall and roof systems are wood shingle

multi-pane window above single pane

roofs tend to be irregular, steeply pitched and typically have cross gables

eaves are at different levels

no corner boards

porch supports are unadorned - shingled, stone or simple wood columns

asymmetrical facade

palladian window

0356

Tudor
1900 to 1940's

windows are typically casement or double-hung

decorative half-timbering is common

half-timbering infill is usually stucco but can be decorative brick

steep side gabled roof with front facing cross gable(s)

emphasis on roofs and chimneys

overlapping gables

large chimney with decorative pot

entrance doors often have rounded tops

tall, narrow windows that are often leaded and usually arranged in groups

0357

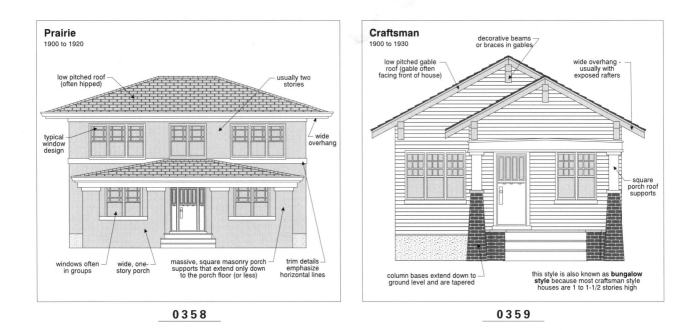

Prairie
1900 to 1920

low pitched roof
(often hipped)

usually two
stories

typical
window
design

wide
overhang

windows often
in groups

wide, one-
story porch

massive, square masonry porch
supports that extend only down
to the porch floor (or less)

trim details
emphasize
horizontal lines

0358

Craftsman
1900 to 1930

decorative beams
or braces in gables

low pitched gable
roof (gable often
facing front of house)

wide overhang -
usually with
exposed rafters

square
porch roof
supports

column bases extend down to
ground level and are tapered

this style is also known as **bungalow
style** because most craftsman style
houses are 1 to 1-1/2 stories high

0359

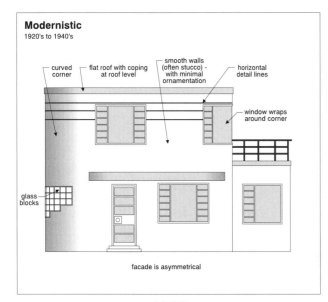

Modernistic
1920's to 1940's

curved
corner

flat roof with coping
at roof level

smooth walls
(often stucco) -
with minimal
ornamentation

horizontal
detail lines

window wraps
around corner

glass
blocks

facade is asymmetrical

0360

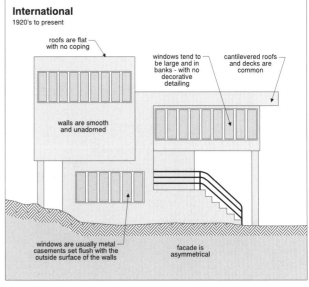

International
1920's to present

roofs are flat
with no coping

windows tend to
be large and in
banks - with no
decorative
detailing

cantilevered roofs
and decks are
common

walls are smooth
and unadorned

windows are usually metal
casements set flush with the
outside surface of the walls

facade is
asymmetrical

0361

Spanish Colonial
1600's to late 1800's

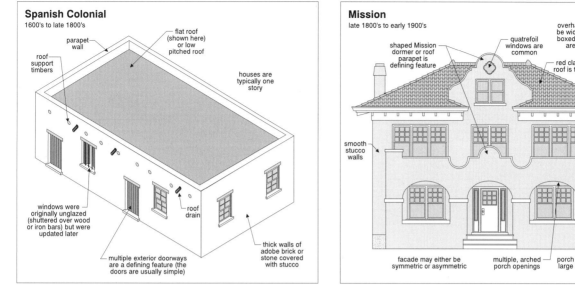

parapet wall

roof support timbers

flat roof (shown here) or low pitched roof

houses are typically one story

windows were originally unglazed (shuttered over wood or iron bars) but were updated later

roof drain

multiple exterior doorways are a defining feature (the doors are usually simple)

thick walls of adobe brick or stone covered with stucco

0362

Mission
late 1800's to early 1900's

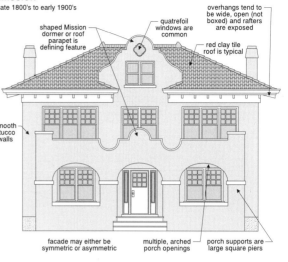

shaped Mission dormer or roof parapet is defining feature

quatrefoil windows are common

overhangs tend to be wide, open (not boxed) and rafters are exposed

red clay tile roof is typical

smooth stucco walls

facade may either be symmetric or asymmetric

multiple, arched porch openings

porch supports are large square piers

0363

Cape Cod
1700's onwards

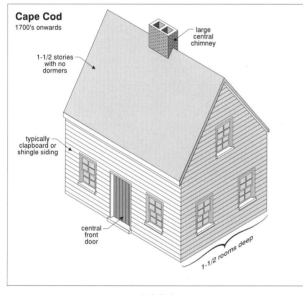

large central chimney

1-1/2 stories with no dormers

typically clapboard or shingle siding

central front door

1-1/2 rooms deep

0364

Saltbox
1700 to 1750's

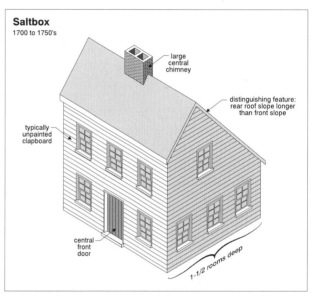

large central chimney

distinguishing feature: rear roof slope longer than front slope

typically unpainted clapboard

central front door

1-1/2 rooms deep

0365

Exterior inspection includes other systems

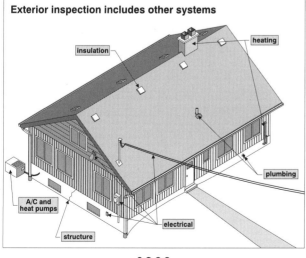

insulation

heating

plumbing

electrical

A/C and heat pumps

structure

0366

Wall assemblies

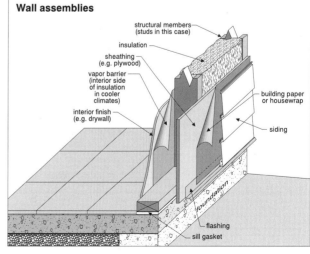

structural members (studs in this case)

insulation

sheathing (e.g. plywood)

vapor barrier (interior side of insulation in cooler climates)

interior finish (e.g. drywall)

building paper or housewrap

siding

foundation

flashing

sill gasket

0367

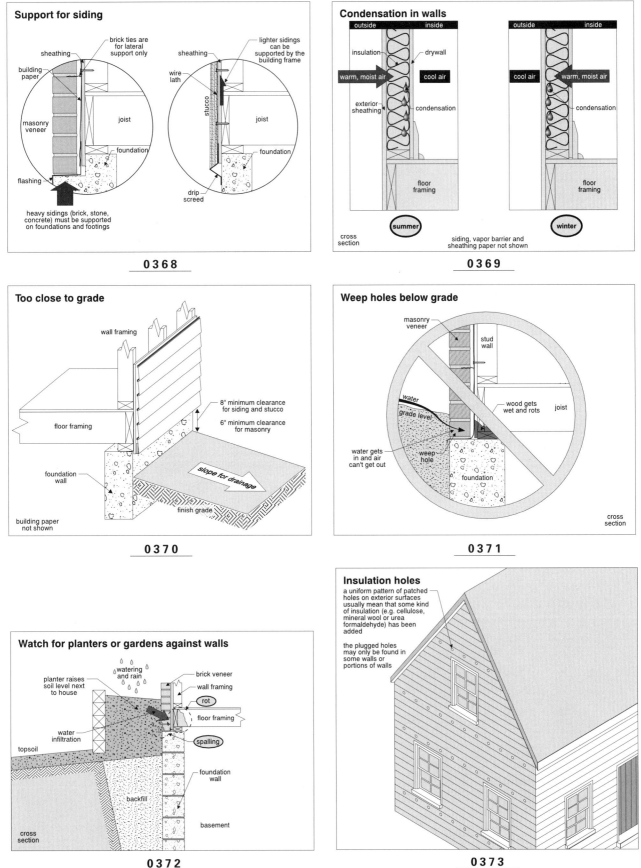

Support for siding

sheathing

building paper

brick ties are for lateral support only

masonry veneer

joist

foundation

flashing

heavy sidings (brick, stone, concrete) must be supported on foundations and footings

sheathing

lighter sidings can be supported by the building frame

wire lath

stucco

joist

foundation

drip screed

0368

Condensation in walls

outside   inside

insulation   drywall

warm, moist air   cool air

exterior sheathing   condensation

floor framing

summer

outside   inside

cool air   warm, moist air

condensation

floor framing

winter

cross section

siding, vapor barrier and sheathing paper not shown

0369

Too close to grade

wall framing

floor framing

8" minimum clearance for siding and stucco

6" minimum clearance for masonry

foundation wall

slope for drainage

finish grade

building paper not shown

0370

Weep holes below grade

masonry veneer

stud wall

water

grade level

wood gets wet and rots   joist

water gets in and air can't get out

weep hole

foundation

cross section

0371

Watch for planters or gardens against walls

watering and rain

planter raises soil level next to house

brick veneer

wall framing

rot

floor framing

water infiltration

spalling

topsoil

foundation wall

backfill

basement

cross section

0372

Insulation holes

a uniform pattern of patched holes on exterior surfaces usually mean that some kind of insulation (e.g. cellulose, mineral wool or urea formaldehyde) has been added

the plugged holes may only be found in some walls or portions of walls

0373

Veneer versus solid masonry

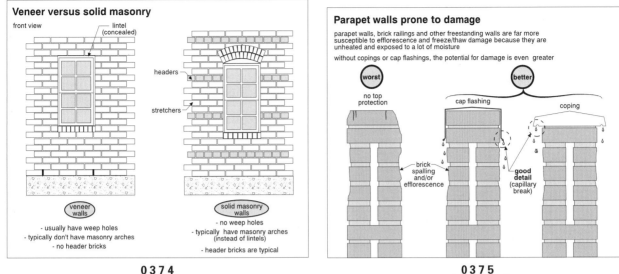

front view

lintel (concealed)

headers

stretchers

veneer walls
- usually have weep holes
- typically don't have masonry arches
- no header bricks

solid masonry walls
- no weep holes
- typically have masonry arches (instead of lintels)
- header bricks are typical

0374

Parapet walls prone to damage

parapet walls, brick railings and other freestanding walls are far more susceptible to efflorescence and freeze/thaw damage because they are unheated and exposed to a lot of moisture

without copings or cap flashings, the potential for damage is even greater

worst

better

no top protection

cap flashing

coping

brick spalling and/or efflorescence

good detail (capillary break)

0375

Chimney deterioration due to condensation

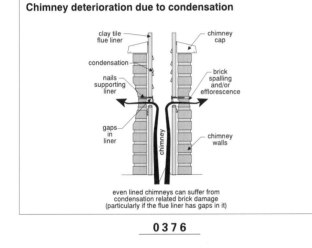

clay tile flue liner

chimney cap

condensation

nails supporting liner

brick spalling and/or efflorescence

gaps in liner

chimney walls

chimney

even lined chimneys can suffer from condensation related brick damage (particularly if the flue liner has gaps in it)

0376

Rising damp

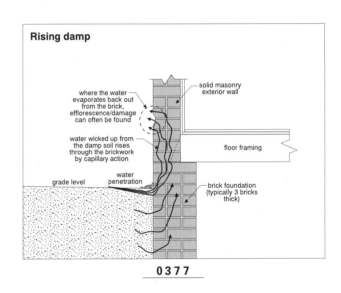

where the water evaporates back out from the brick, efflorescence/damage can often be found

solid masonry exterior wall

water wicked up from the damp soil rises through the brickwork by capillary action

floor framing

grade level

water penetration

brick foundation (typically 3 bricks thick)

0377

Freezing water spalls bricks

freezing temperatures, saturated brick and a susceptible type of brick are required for spalling to occur

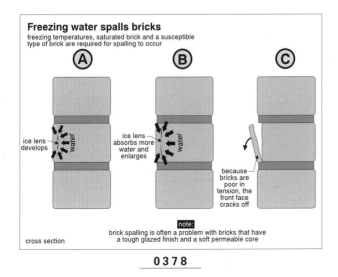

(A)   (B)   (C)

ice lens develops

water

ice lens absorbs more water and enlarges

water

because bricks are poor in tension, the front face cracks off

note: brick spalling is often a problem with bricks that have a tough glazed finish and a soft permeable core

cross section

0378

Repointing

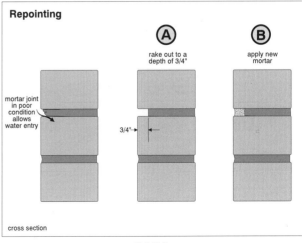

(A)   (B)

rake out to a depth of 3/4"

apply new mortar

mortar joint in poor condition allows water entry

3/4"

cross section

0379

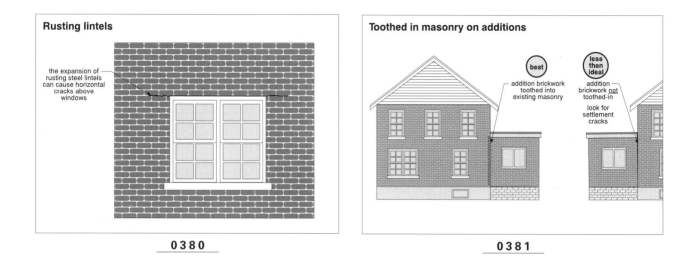

Rusting lintels

the expansion of rusting steel lintels can cause horizontal cracks above windows

0380

Toothed in masonry on additions

best

less than ideal

addition brickwork toothed into existing masonry

addition brickwork not toothed-in

look for settlement cracks

0381

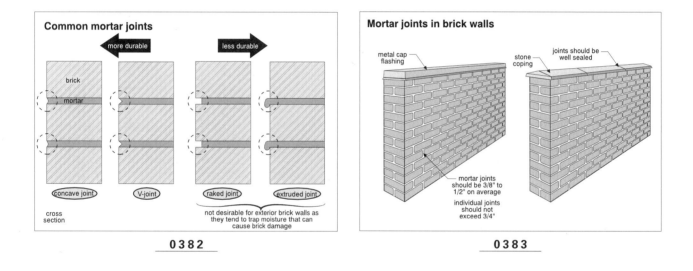

Common mortar joints

more durable

less durable

brick

mortar

concave joint

V-joint

raked joint

extruded joint

cross section

not desirable for exterior brick walls as they tend to trap moisture that can cause brick damage

0382

Mortar joints in brick walls

metal cap flashing

stone coping

joints should be well sealed

mortar joints should be 3/8" to 1/2" on average

individual joints should not exceed 3/4"

0383

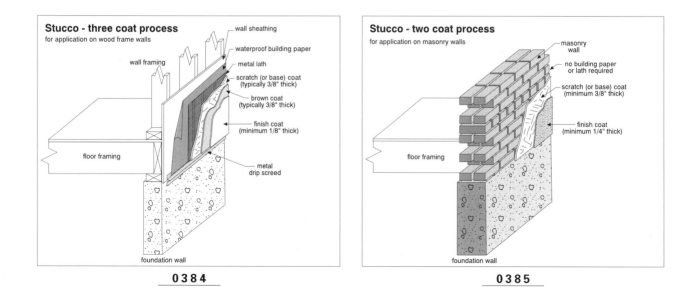

Stucco - three coat process
for application on wood frame walls

wall framing

wall sheathing

waterproof building paper

metal lath

scratch (or base) coat (typically 3/8" thick)

brown coat (typically 3/8" thick)

finish coat (minimum 1/8" thick)

metal drip screed

floor framing

foundation wall

0384

Stucco - two coat process
for application on masonry walls

masonry wall

no building paper or lath required

scratch (or base) coat (minimum 3/8" thick)

finish coat (minimum 1/4" thick)

floor framing

foundation wall

0385

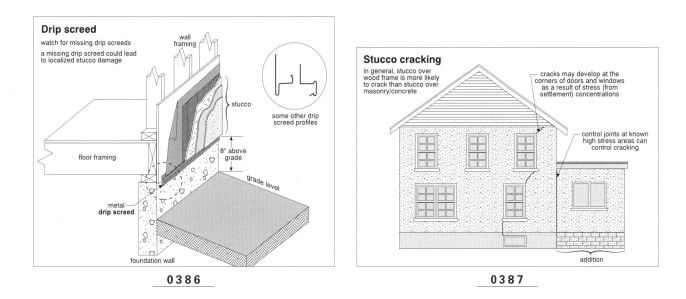

Drip screed

watch for missing drip screeds

a missing drip screed could lead to localized stucco damage

wall framing

stucco

some other drip screed profiles

floor framing

8" above grade

grade level

metal **drip screed**

foundation wall

0386

Stucco cracking

in general, stucco over wood frame is more likely to crack than stucco over masonry/concrete

cracks may develop at the corners of doors and windows as a result of stress (from settlement) concentrations

control joints at known high stress areas can control cracking

addition

0387

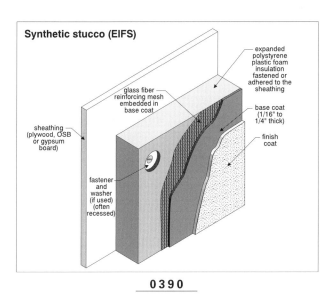

Tudor style stucco problems

stucco

half-timbering

water can collect in these areas leading to localized stucco and wood deterioration

0388

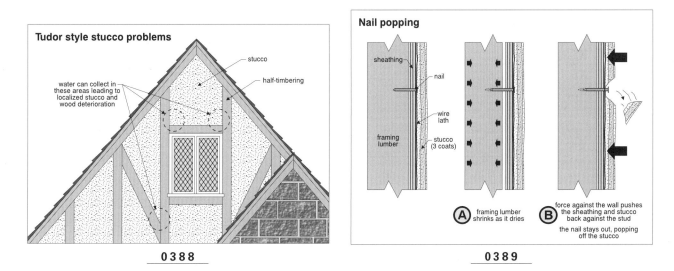

Nail popping

sheathing

nail

wire lath

framing lumber

stucco (3 coats)

(A) framing lumber shrinks as it dries

(B) force against the wall pushes the sheathing and stucco back against the stud

the nail stays out, popping off the stucco

0389

Synthetic stucco (EIFS)

expanded polystyrene plastic foam insulation fastened or adhered to the sheathing

glass fiber reinforcing mesh embedded in base coat

base coat (1/16" to 1/4" thick)

finish coat

sheathing (plywood, OSB or gypsum board)

fastener and washer (if used) (often recessed)

0390

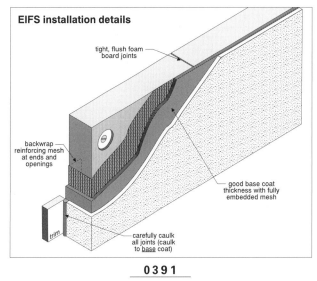

EIFS installation details

tight, flush foam board joints

backwrap reinforcing mesh at ends and openings

good base coat thickness with fully embedded mesh

trim

carefully caulk all joints (caulk to base coat)

0391

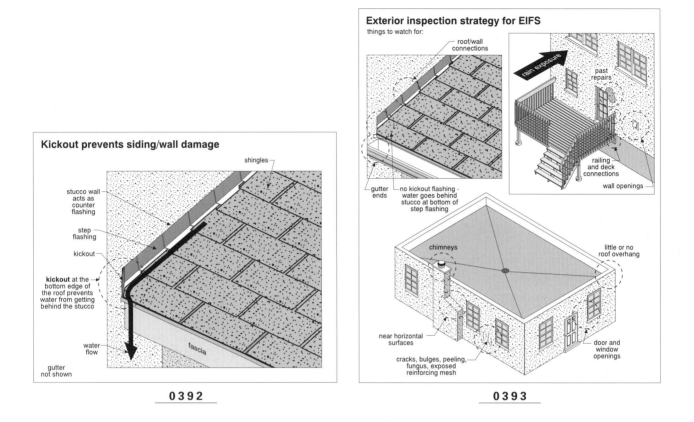

Kickout prevents siding/wall damage

shingles

stucco wall acts as counter flashing

step flashing

kickout

kickout at the bottom edge of the roof prevents water from getting behind the stucco

water flow

fascia

gutter not shown

0392

Exterior inspection strategy for EIFS

things to watch for:

roof/wall connections

rain exposure

past repairs

railing and deck connections

wall openings

gutter ends

no kickout flashing - water goes behind stucco at bottom of step flashing

chimneys

little or no roof overhang

near horizontal surfaces

cracks, bulges, peeling, fungus, exposed reinforcing mesh

door and window openings

0393

Interior inspection strategy for EIFS

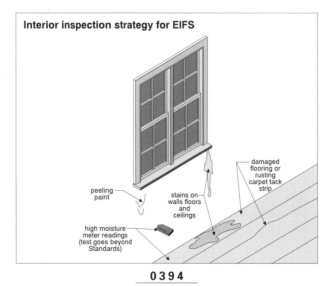

peeling paint

stains on walls floors and ceilings

high moisture meter readings (test goes beyond Standards)

damaged flooring or rusting carpet tack strip

0394

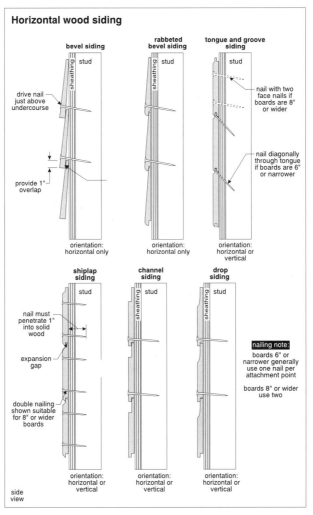

Horizontal wood siding

bevel siding

stud

sheathing

drive nail just above undercourse

provide 1" overlap

orientation: horizontal only

rabbeted bevel siding

stud

sheathing

orientation: horizontal only

tongue and groove siding

stud

nail with two face nails if boards are 8" or wider

nail diagonally through tongue if boards are 6" or narrower

orientation: horizontal or vertical

shiplap siding

stud

nail must penetrate 1" into solid wood

expansion gap

double nailing shown suitable for 8" or wider boards

orientation: horizontal or vertical

channel siding

stud

sheathing

orientation: horizontal or vertical

drop siding

stud

sheathing

nailing note:
boards 6" or narrower generally use one nail per attachment point

boards 8" or wider use two

orientation: horizontal or vertical

side view

0395

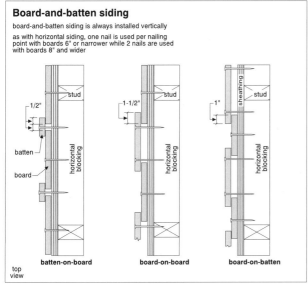

Board-and-batten siding

board-and-batten siding is always installed vertically

as with horizontal siding, one nail is used per nailing point with boards 6" or narrower while 2 nails are used with boards 8" and wider

stud

1/2"

batten

board

horizontal blocking

batten-on-board

1-1/2"

stud

horizontal blocking

board-on-board

sheathing

stud

1"

horizontal blocking

board-on-batten

top view

0396

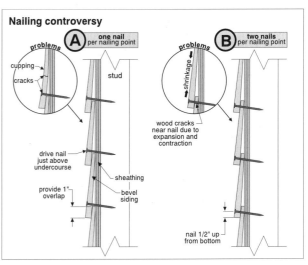

Nailing controversy

(A) one nail per nailing point

problems

cupping

cracks

stud

drive nail just above undercourse

provide 1" overlap

sheathing

bevel siding

(B) two nails per nailing point

problems

shrinkage

wood cracks near nail due to expansion and contraction

nail 1/2" up from bottom

0397

Concealed nails

narrow siding may have one <u>concealed</u> nail per nailing point

siding nails that have rounded heads are often used to keep the boards slightly separated - allowing air in behind the siding to dry the back of the boards

stud

air

sheathing

narrow siding

0398

Inside and outside corners

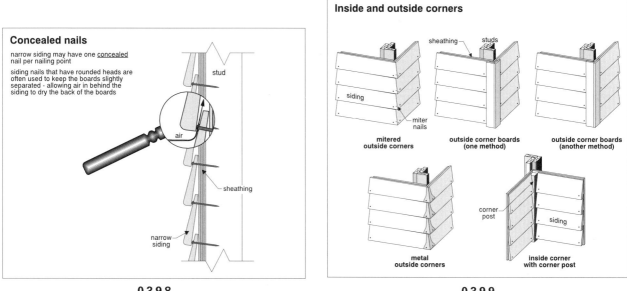

sheathing   studs

siding

miter nails

mitered outside corners

outside corner boards (one method)

outside corner boards (another method)

corner post

siding

metal outside corners

inside corner with corner post

0399

Wood shingles and shakes

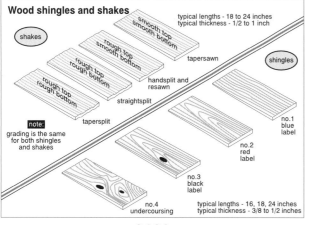

shakes

typical lengths - 18 to 24 inches
typical thickness - 1/2 to 1 inch

smooth top smooth bottom

tapersawn

rough top smooth bottom

handsplit and resawn

rough top rough bottom

straightsplit

rough top rough bottom

tapersplit

shingles

no.1 blue label

no.2 red label

note:
grading is the same for both shingles and shakes

no.3 black label

no.4 undercoursing

typical lengths - 16, 18, 24 inches
typical thickness - 3/8 to 1/2 inches

0400

Cedar shingles - nailing details

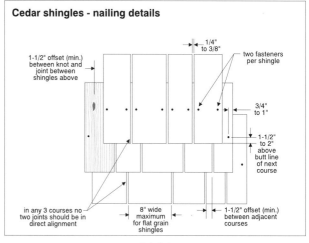

1/4" to 3/8"

two fasteners per shingle

1-1/2" offset (min.) between knot and joint between shingles above

3/4" to 1"

1-1/2" to 2" above butt line of next course

in any 3 courses no two joints should be in direct alignment

8" wide maximum for flat grain shingles

1-1/2" offset (min.) between adjacent courses

0401

Wood shingle siding

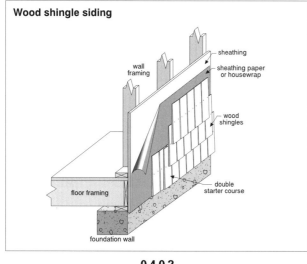

wall framing

sheathing

sheathing paper or housewrap

wood shingles

floor framing

double starter course

foundation wall

0402

Cut back bottom of siding

cutting the bottom of the siding back at 45° reduces the amount of water which can soak into the end grain and suggests good workmanship

siding

joist

foundation

45°

0403

Mitered corners

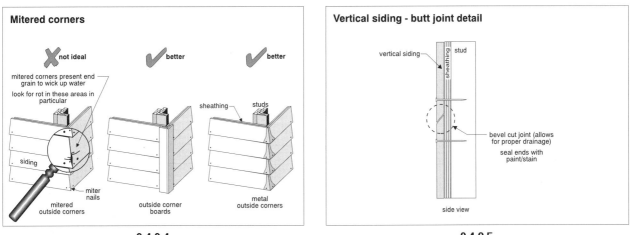

not ideal

mitered corners present end grain to wick up water

look for rot in these areas in particular

siding

miter nails

mitered outside corners

better

outside corner boards

better

sheathing   studs

metal outside corners

0404

Vertical siding - butt joint detail

vertical siding

sheathing

stud

bevel cut joint (allows for proper drainage)

seal ends with paint/stain

side view

0405

Joints in siding boards

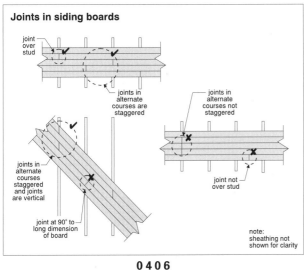

joint over stud

joints in alternate courses are staggered

joints in alternate courses not staggered

joints in alternate courses staggered and joints are vertical

joint at 90° to long dimension of board

joint not over stud

note: sheathing not shown for clarity

0406

Strapping for vertical siding

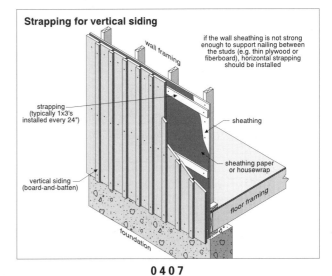

wall framing

if the wall sheathing is not strong enough to support nailing between the studs (e.g. thin plywood or fiberboard), horizontal strapping should be installed

strapping (typically 1x3's installed every 24")

sheathing

sheathing paper or housewrap

vertical siding (board-and-batten)

floor framing

foundation

0407

Plywood and composite siding

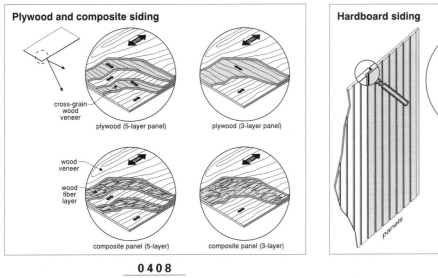

cross-grain wood veneer

plywood (5-layer panel)

plywood (3-layer panel)

wood veneer

wood fiber layer

composite panel (5-layer)

composite panel (3-layer)

0408

Hardboard siding

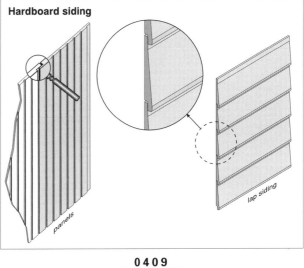

panels

lap siding

0409

Oriented strandboard (OSB) siding

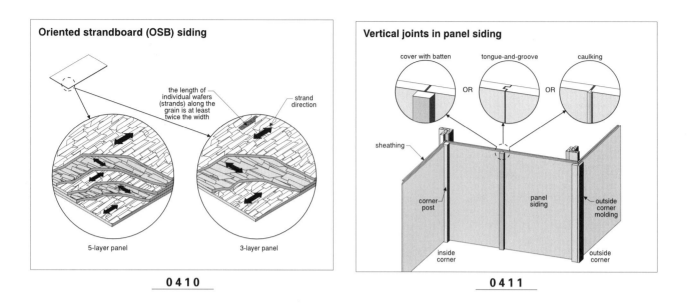

the length of individual wafers (strands) along the grain is at least twice the width

strand direction

5-layer panel

3-layer panel

0 4 1 0

Vertical joints in panel siding

cover with batten

tongue-and-groove

caulking

OR

OR

sheathing

corner post

panel siding

outside corner molding

inside corner

outside corner

0 4 1 1

Siding too close to roof

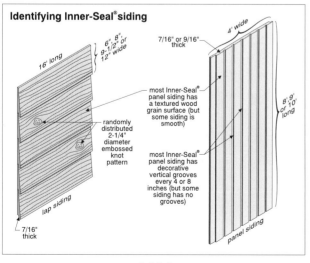

siding should have 1" (preferably 2") clearance from roof shingles to prevent water damage

metal step flashings

0 4 1 2

Z flashings

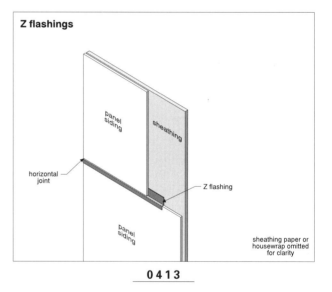

panel siding

sheathing

horizontal joint

Z flashing

panel siding

sheathing paper or housewrap omitted for clarity

0 4 1 3

Identifying Inner-Seal® siding

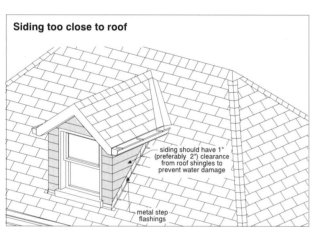

16' long

6", 8" or 9-1/2" or 12" wide

4' wide

7/16" or 9/16" thick

most Inner-Seal® panel siding has a textured wood grain surface (but some siding is smooth)

8', 9' or 10' long

randomly distributed 2-1/4" diameter embossed knot pattern

most Inner-Seal® panel siding has decorative vertical grooves every 4 or 8 inches (but some siding has no grooves)

lap siding

7/16" thick

panel siding

0 4 1 4

Metal and vinyl siding

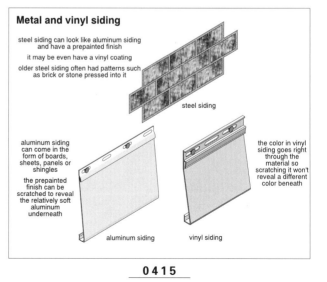

steel siding can look like aluminum siding and have a prepainted finish

it may be even have a vinyl coating

older steel siding often had patterns such as brick or stone pressed into it

steel siding

aluminum siding can come in the form of boards, sheets, panels or shingles

the prepainted finish can be scratched to reveal the relatively soft aluminum underneath

the color in vinyl siding goes right through the material so scratching it won't reveal a different color beneath

aluminum siding

vinyl siding

0 4 1 5

Nailing too tightly

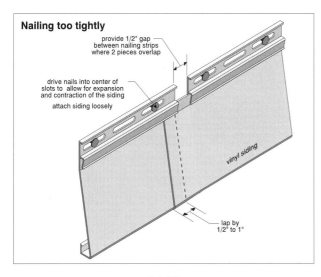

provide 1/2" gap between nailing strips where 2 pieces overlap

drive nails into center of slots to allow for expansion and contraction of the siding

attach siding loosely

vinyl siding

lap by 1/2" to 1"

0 4 1 6

Mounting blocks

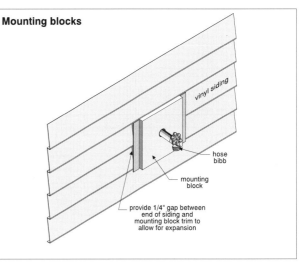

vinyl siding

hose bibb

mounting block

provide 1/4" gap between end of siding and mounting block trim to allow for expansion

0 4 1 7

Flashings below windows

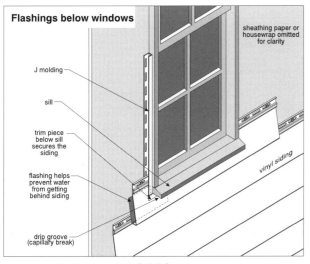

sheathing paper or housewrap omitted for clarity

J molding

sill

trim piece below sill secures the siding

flashing helps prevent water from getting behind siding

vinyl siding

drip groove (capillary break)

0 4 1 8

J moldings around windows

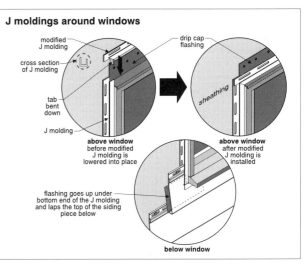

modified J molding

cross section of J molding

tab bent down

J molding

drip cap flashing

sheathing

above window
before modified J molding is lowered into place

above window
after modified J molding is installed

flashing goes up under bottom end of the J molding and laps the top of the siding piece below

below window

0 4 1 9

Asbestos cement siding

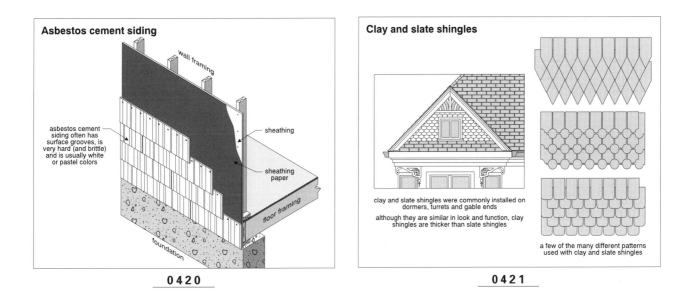

wall framing

asbestos cement siding often has surface grooves, is very hard (and brittle) and is usually white or pastel colors

sheathing

sheathing paper

floor framing

foundation

0 4 2 0

Clay and slate shingles

clay and slate shingles were commonly installed on dormers, turrets and gable ends

although they are similar in look and function, clay shingles are thicker than slate shingles

a few of the many different patterns used with clay and slate shingles

0 4 2 1

Six nails per shingle

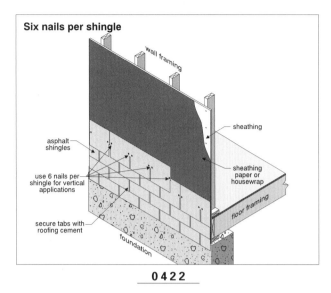

wall framing

sheathing

asphalt shingles

sheathing paper or housewrap

use 6 nails per shingle for vertical applications

secure tabs with roofing cement

floor framing

foundation

0422

Insulbrick

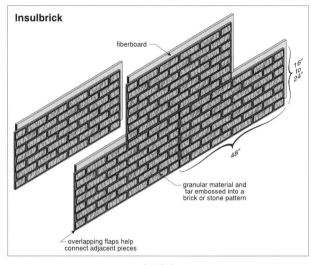

fiberboard

16" to 24"

48"

granular material and tar embossed into a brick or stone pattern

overlapping flaps help connect adjacent pieces

0423

Foundation height

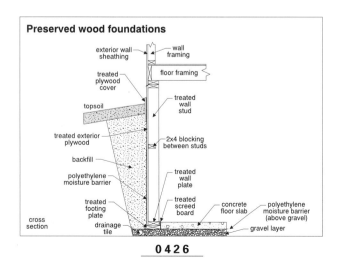

wall framing

wood siding

bottom of siding

8"

grade level

floor framing

top of foundation

10"

0424

Areas to check for wood/soil contact include:

retaining wall

0425

Preserved wood foundations

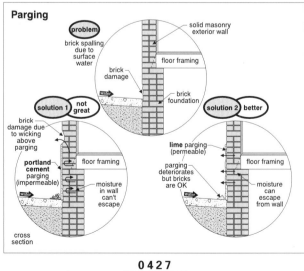

exterior wall sheathing

wall framing

treated plywood cover

floor framing

topsoil

treated wall stud

treated exterior plywood

backfill

2x4 blocking between studs

polyethylene moisture barrier

treated wall plate

treated footing plate

treated screed board

concrete floor slab

polyethylene moisture barrier (above gravel)

cross section

drainage tile

gravel layer

0426

Parging

problem

brick spalling due to surface water

solid masonry exterior wall

floor framing

brick damage

brick foundation

water

solution 1 **not great**

brick damage due to wicking above parging

portland cement parging (impermeable)

floor framing

moisture in wall can't escape

water

solution 2 **better**

lime parging (permeable)

parging deteriorates but bricks are OK

floor framing

moisture can escape from wall

water

cross section

0427

Soffits and fascia

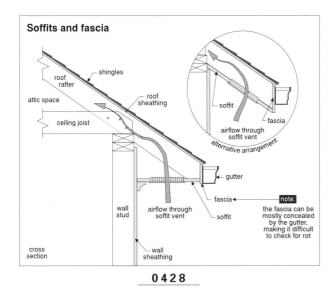

shingles

roof rafter

attic space

roof sheathing

ceiling joist

soffit

fascia

airflow through soffit vent

alternative arrangement

gutter

fascia

soffit

wall stud

airflow through soffit vent

note:
the fascia can be mostly concealed by the gutter, making it difficult to check for rot

wall sheathing

cross section

0428

Rotting soffit and fascia

the circled areas warrant special attention as they are particularly prone to soffit and fascia rot

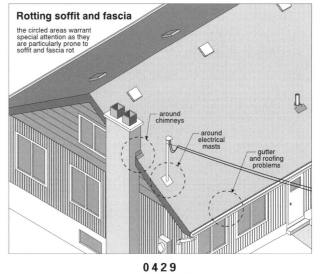

around chimneys

around electrical masts

gutter and roofing problems

0429

Putty (glazing compound)

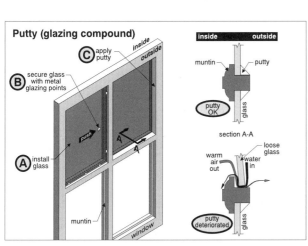

apply putty

secure glass with metal glazing points

install glass

push

muntin

inside   outside

muntin

putty

putty OK

section A-A

warm air out

loose glass

water in

putty deteriorated

glass

window

0430

Capillary break

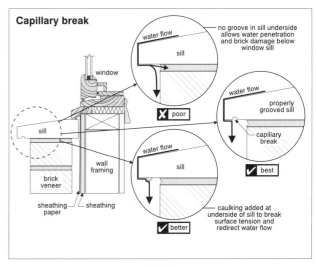

water flow

no groove in sill underside allows water penetration and brick damage below window sill

sill

window

water flow

properly grooved sill

sill

capillary break

sill

X poor

✔ best

wall framing

brick veneer

water flow

sill

sheathing paper

sheathing

✔ better

caulking added at underside of sill to break surface tension and redirect water flow

0431

Windows move

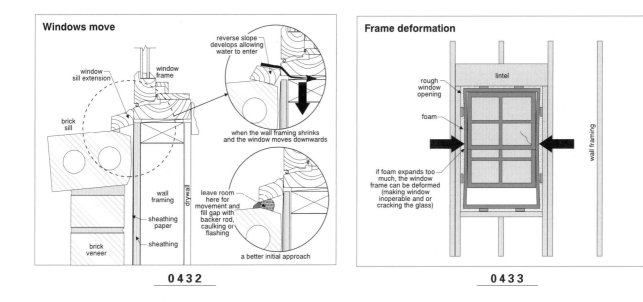

window sill extension

window frame

reverse slope develops allowing water to enter

brick sill

when the wall framing shrinks and the window moves downwards

wall framing

drywall

leave room here for movement and fill gap with backer rod, caulking or flashing

sheathing paper

sheathing

brick veneer

a better initial approach

0432

Frame deformation

lintel

rough window opening

foam

wall framing

if foam expands too much, the window frame can be deformed (making window inoperable and or cracking the glass)

0433

Flashings over windows

note:
a drip cap flashing is not required if the roof overhang width is four or more times greater than the distance from the top of the window to the soffit

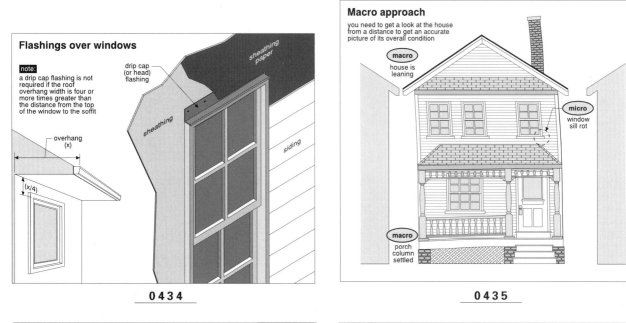

drip cap (or head) flashing

sheathing paper

sheathing

siding

overhang (x)

(x/4)

0434

Macro approach

you need to get a look at the house from a distance to get an accurate picture of its overall condition

macro
house is leaning

micro
window sill rot

macro
porch column settled

0435

Porches

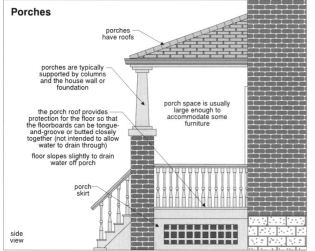

porches have roofs

porches are typically supported by columns and the house wall or foundation

the porch roof provides protection for the floor so that the floorboards can be tongue-and-groove or butted closely together (not intended to allow water to drain through)

floor slopes slightly to drain water off porch

porch skirt

porch space is usually large enough to accommodate some furniture

side view

0436

Decks

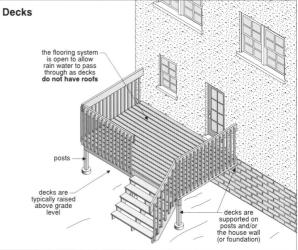

the flooring system is open to allow rain water to pass through as decks **do not have roofs**

posts

decks are typically raised above grade level

decks are supported on posts and/or the house wall (or foundation)

0437

Balconies

the defining feature of balconies is their lack of steps and access to grade level

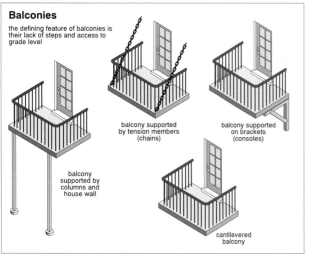

balcony supported by tension members (chains)

balcony supported on brackets (consoles)

balcony supported by columns and house wall

cantilevered balcony

0438

Basement walkout

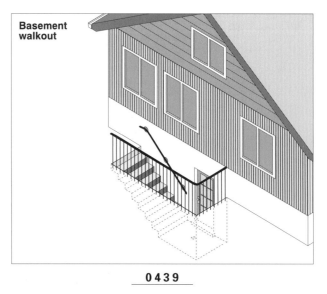

0439

Support for concrete steps

floating

supported on their own foundation (continuous)

supported by building foundation (cantilevered)

supported on their own foundation (pier)

side view

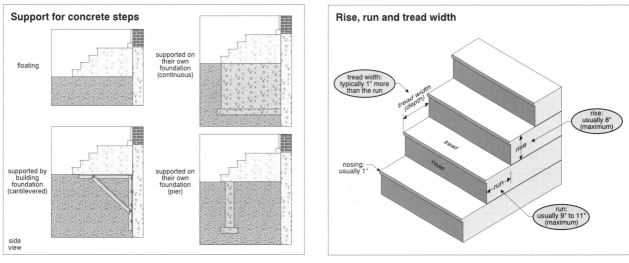

0440

Rise, run and tread width

tread width: typically 1" more than the run

tread width (depth)

rise: usually 8" (maximum)

tread

rise

riser

nosing: usually 1"

run

run: usually 9" to 11" (maximum)

0441

Landings

a minimum 3' by 3' landing should be provided in front of an entrance door

in some areas, a landing is not required for secondary entrances that are served by stairs with three or less risers

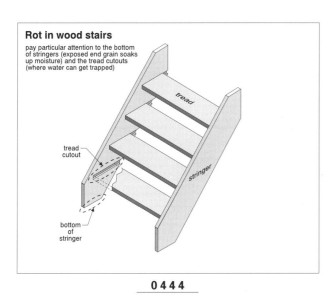

0442

Settled or heaved steps

riser height not uniform

gap (wider at top)

treads and landing out of level

treads out of level

settlement

steps and landing settled

settlement

steps settled away from landing

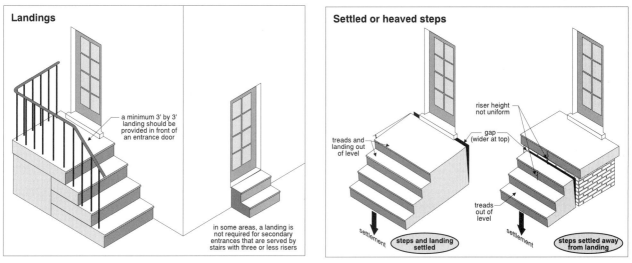

0443

Rot in wood stairs

pay particular attention to the bottom of stringers (exposed end grain soaks up moisture) and the tread cutouts (where water can get trapped)

tread

tread cutout

stringer

bottom of stringer

0444

Designing steps that aren't springy

1"

front of tread is supported by continuous riser

tread

riser

1-1/2"

front of tread is unsupporte

tread

stringer

3-1/2" minimum

9-1/2" minimum

1-1/2"

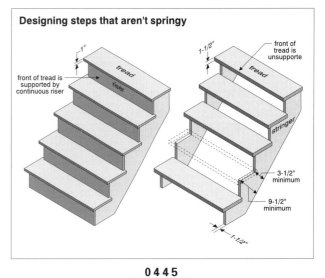

0445

Handrails and guards

guards

guards are required if the floor of the deck, porch or balcony is more than <u>24" to 30"</u> above grade

height above grade

1 handrail required: if more than 3 risers

hand-rails

2 handrails required: if more than 3 risers and stair width >44"

stair width

0446

Handrail design

1-1/2" 1-1/2"

top of handrail should be easy to grip (tube or oval)

wall

A A

section A-A

0447

Spindle spacing

spindles should be spaced so that a 4" (6" in some areas) sphere cannot pass through the guard

spindles (balusters)

4" diameter sphere (6" in some jurisdictions)

construction note:

front view

horizontal details that make climbing the guard easier should be avoided

0448

Handrails or guards too low

42"

balcony or deck more than 6' above grade

42" in some areas regardless of deck height

36"

handrail height (H) should be between 31" and 34"

note:
in some jurisdictions H should be between 34" and 38"

H

line through nosing

balcony, deck or landing less than 6' above grade

0449

Common column materials

hollow wood (>6" diameter)

wood surrounding steel column

A A

solid wood (<6" diameter)

column

masonry

metal

section A-A

side view

0450

Column hinge point

the dissimilar materials create a natural hinge point here

plumb line

the weight of the porch roof is an eccentric load that will try to push out the top of the masonry portion of the

side view

0451

Beams

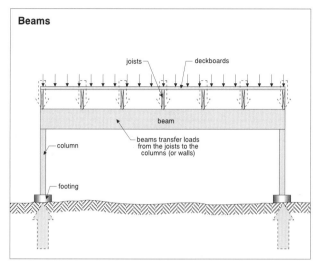

joists

deckboards

beam

beams transfer loads from the joists to the columns (or walls)

column

footing

0452

Inspecting for beam sag

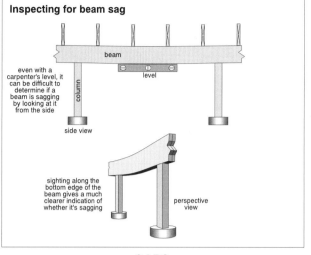

beam

level

column

even with a carpenter's level, it can be difficult to determine if a beam is sagging by looking at it from the side

side view

sighting along the bottom edge of the beam gives a much clearer indication of whether it's sagging

perspective view

0453

Poor end support

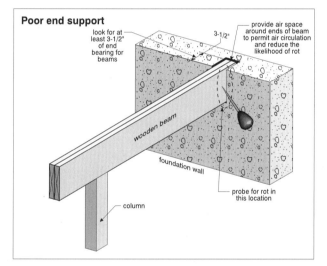

look for at least 3-1/2" of end bearing for beams

provide air space around ends of beam to permit air circulation and reduce the likelihood of rot

3-1/2"

wooden beam

foundation wall

probe for rot in this location

column

0454

Beam rotation (twisting)

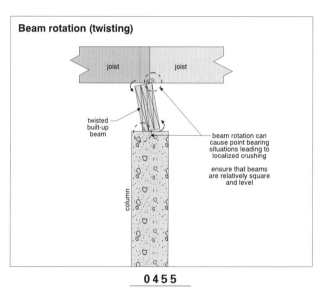

joist

joist

twisted built-up beam

beam rotation can cause point bearing situations leading to localized crushing

ensure that beams are relatively square and level

column

0455

Joist support

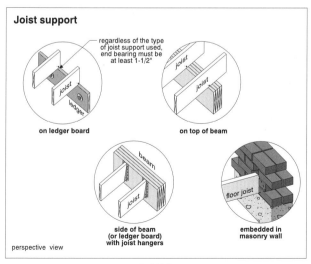

regardless of the type of joist support used, end bearing must be at least 1-1/2"

joist

ledger

joist

on ledger board

on top of beam

beam

joist

floor joist

side of beam (or ledger board) with joist hangers

embedded in masonry wall

perspective view

0456

Porch floor sag

drainage

floor boards (tongue-and-groove)

porch columns

joists

makeshift intermediate support (added after joists sagged)

long joist span so that floor boards can be installed perpendicular to the house for drainage

beam

porch roof and supports not shown

0457

Securing ledgerboards

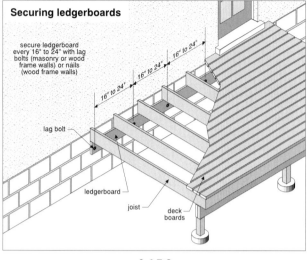

secure ledgerboard every 16" to 24" with lag bolts (masonry or wood frame walls) or nails (wood frame walls)

16" to 24"

lag bolt

ledgerboard

joist

deck boards

0458

Ledgerboard flashing

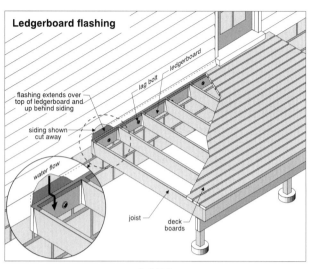

flashing extends over top of ledgerboard and up behind siding

siding shown cut away

water flow

lag bolt

ledgerboard

joist

deck boards

0459

Cantilevered decks

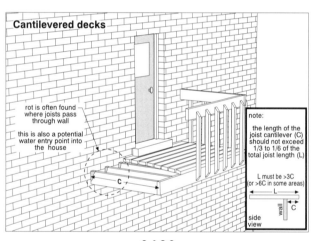

rot is often found where joists pass through wall

this is also a potential water entry point into the house

C

note:

the length of the joist cantilever (C) should not exceed 1/3 to 1/6 of the total joist length (L)

L must be >3C (or >6C in some areas)

L

C

wall

side view

0460

Canvas duck

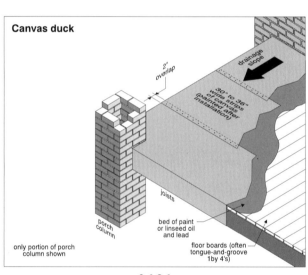

2" overlap

drainage slope

30" to 36" wide strips of canvas (painted after installation)

joists

porch column

only portion of porch column shown

bed of paint or linseed oil and lead

floor boards (often tongue-and-groove 1 by 4's)

0461

Deck board installation details

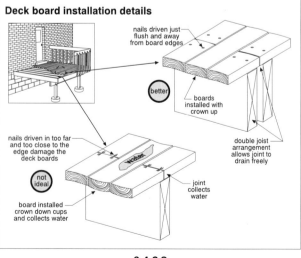

nails driven just flush and away from board edges

better

boards installed with crown up

nails driven in too far and too close to the edge damage the deck boards

not ideal

water

board installed crown down cups and collects water

joint collects water

double joist arrangement allows joint to drain freely

0462

Threshold

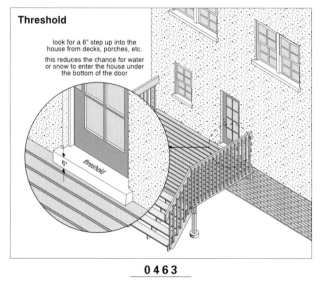

look for a 6" step up into the house from decks, porches, etc.

this reduces the chance for water or snow to enter the house under the bottom of the door

6"

threshold

0463

Hinged pillars affect roof structure

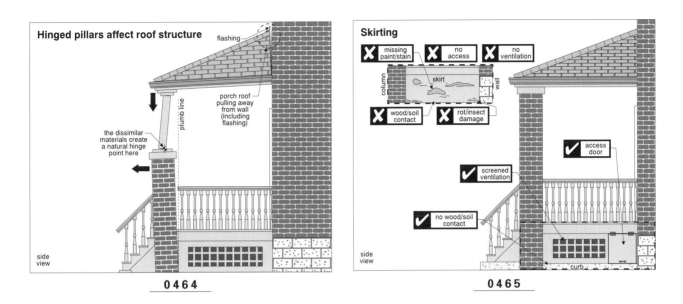

flashing

porch roof pulling away from wall (including flashing)

plumb line

the dissimilar materials create a natural hinge point here

side view

0464

Skirting

missing paint/stain — no access — no ventilation

column — skirt — wall

wood/soil contact — rot/insect damage

access door

screened ventilation

no wood/soil contact

curb

side view

0465

Garages versus carports

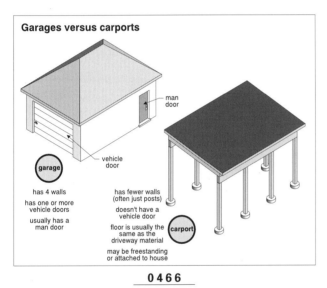

man door

vehicle door

garage

has 4 walls

has one or more vehicle doors

usually has a man door

has fewer walls (often just posts)

doesn't have a vehicle door

floor is usually the same as the driveway material

may be freestanding or attached to house

carport

0466

Fire and gas proofing in attached garages

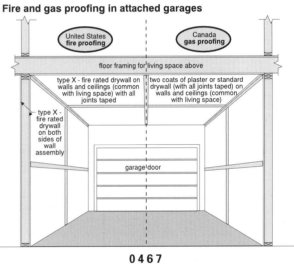

United States fire proofing — Canada gas proofing

floor framing for living space above

type X - fire rated drywall on walls and ceilings (common with living space) with all joints taped

two coats of plaster or standard drywall (with all joints taped) on walls and ceilings (common with living space)

type X - fire rated drywall on both sides of wall assembly

garage door

0467

Heating ductwork in garages

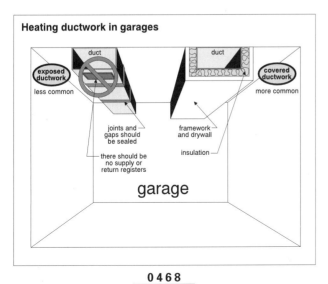

duct — duct

exposed ductwork less common

covered ductwork more common

joints and gaps should be sealed

framework and drywall

there should be no supply or return registers

insulation

garage

0468

Man door (attached garage)

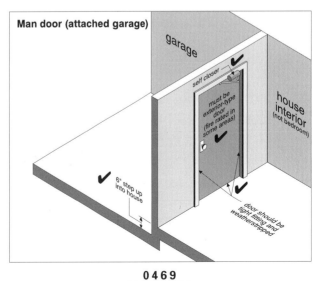

garage

self closer

must be exterior-type door (fire rated in some areas)

house interior (not bedroom)

6" step up into house

door should be tight fitting and weatherstripped

0469

Combustible insulation in garages

✗ exposed plastic insulation on garage door, walls and ceiling	✓ insulation removed or covered with noncombustible material (drywall)

plastic insulation

plastic insulation
plastic insulation
plastic insulation
plastic insulation
plastic insulation

remove plastic insulation

cover plastic insulation on walls and ceilings

0470

Structural garage floor

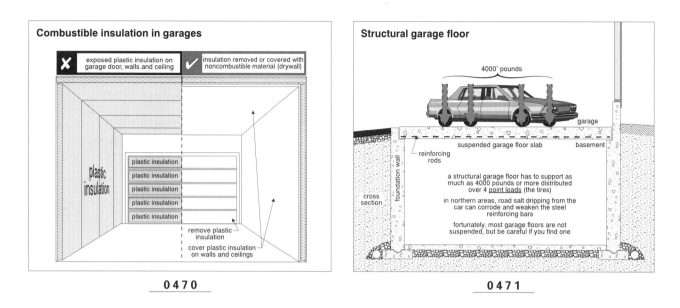

4000⁺ pounds

garage

suspended garage floor slab   basement

foundation wall

reinforcing rods

cross section

a structural garage floor has to support as much as 4000 pounds or more distributed over 4 <u>point loads</u> (the tires)

in northern areas, road salt dripping from the car can corrode and weaken the steel reinforcing bars

fortunately, most garage floors are not suspended, but be careful if you find one

0471

Poor garage floor drainage

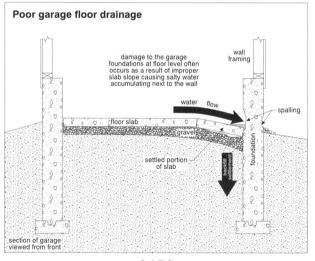

wall framing

damage to the garage foundations at floor level often occurs as a result of improper slab slope causing salty water accumulating next to the wall

water flow

spalling

floor slab

gravel

settled portion of slab

backfill settlement

foundation

section of garage viewed from front

0472

Slab is partly on disturbed soil

garage

wall framing

floor slab settles at edges where backfill is thickest and settles the most

crack   crack

backfill   gravel

backfill settlement

solution: during construction, excavate central mound of soil to at least 3 feet below slab

backfill settlement

foundation

backfill

undisturbed soil

section of garage viewed from front

0473

Mud jacking (pressure grouting)

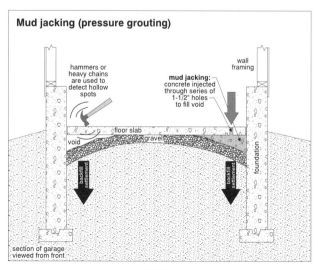

wall framing

hammers or heavy chains are used to detect hollow spots

mud jacking: concrete injected through series of 1-1/2" holes to fill void

floor slab

void   gravel

backfill settlement   backfill settlement

foundation

section of garage viewed from front

0474

Drains for below-grade garages

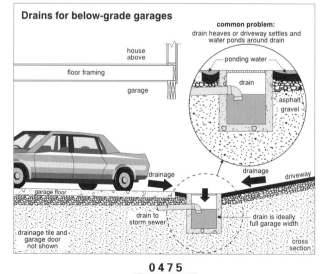

common problem: drain heaves or driveway settles and water ponds around drain

house above

floor framing

garage

ponding water

drain

asphalt

gravel

drainage   drainage   driveway

garage floor

drain to storm sewer

drain is ideally full garage width

drainage tile and garage door not shown

cross section

0475

Drain pipe should face down

check for grates to reduce debris entering catch basin

with straight pipe, debris can easily enter and clog outlet

outlet with elbow is much less likely to clog with debris

outlet pipe

clog

debris

water

outlet pipe (with elbow)

debris

water

pipe diameter should be minimum 3"

garage/driveway drain

garage/driveway drain

(straight pipe)

(elbow)

0 4 7 6

Garage door types

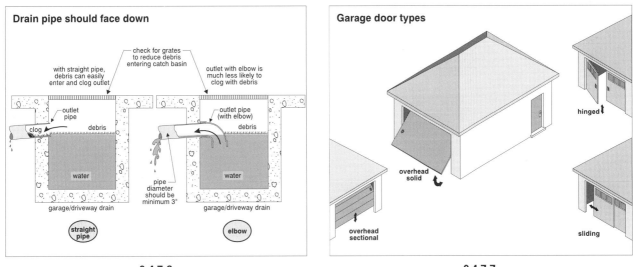

hinged

overhead solid

overhead sectional

sliding

0 4 7 7

Garage door conditions

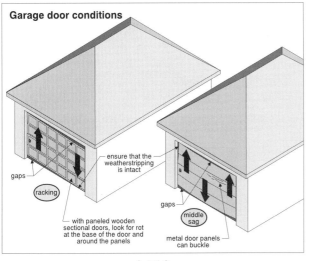

ensure that the weatherstripping is intact

gaps

(racking)

gaps

(middle sag)

with paneled wooden sectional doors, look for rot at the base of the door and around the panels

metal door panels can buckle

0 4 7 8

Automatic garage door openers

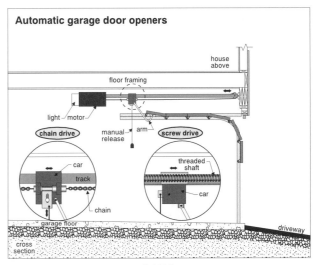

house above

floor framing

light — motor

manual release

arm

(chain drive)

(screw drive)

car

track

threaded shaft

chain

car

garage floor

driveway

cross section

0 4 7 9

Manual operation of automatic garage door openers

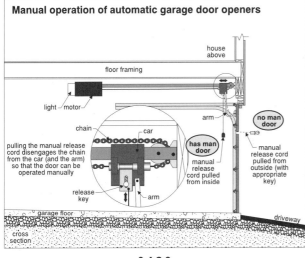

house above

floor framing

light — motor

chain

car

arm

(no man door)

(has man door)

manual release cord pulled from inside

manual release cord pulled from outside (with appropriate key)

pulling the manual release cord disengages the chain from the car (and the arm) so that the door can be operated manually

release key

arm

garage floor

driveway

cross section

0 4 8 0

Check electrical connection

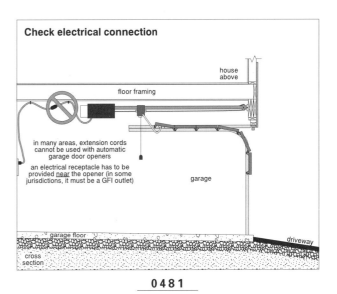

house above

floor framing

in many areas, extension cords cannot be used with automatic garage door openers

an electrical receptacle has to be provided <u>near</u> the opener (in some jurisdictions, it must be a GFI outlet)

garage

garage floor

driveway

cross section

0 4 8 1

Testing automatic reverse

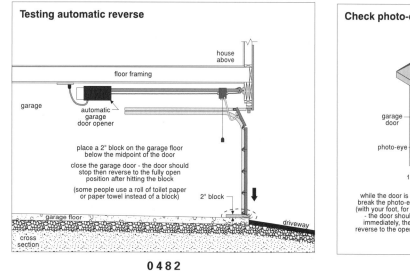

house above

floor framing

garage

automatic garage door opener

place a 2" block on the garage floor below the midpoint of the door

close the garage door - the door should stop then reverse to the fully open position after hitting the block

(some people use a roll of toilet paper or paper towel instead of a block)

2" block

garage floor

driveway

cross section

0482

Check photo-eye

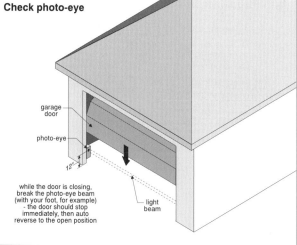

garage door

photo-eye

while the door is closing, break the photo-eye beam (with your foot, for example) - the door should stop immediately, then auto reverse to the open position

light beam

12"

0483

Water heaters in garages

gas or oil fired water heaters that are located in a garage must be at least 18" above the floor - so that gasoline vapors are not ignited by the pilot or burner

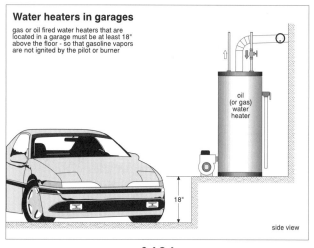

oil (or gas) water heater

18"

side view

0484

Basement walkouts

common problem areas

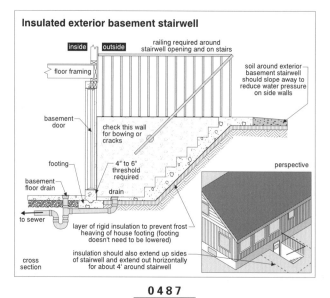

covers/ roofs

steps and railings

door thresholds

drains

frost

walls cracking, leaning, bowing or spalling

0485

Lowered footings around basement walkouts

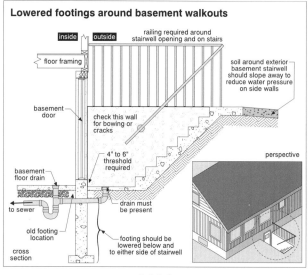

inside | outside

railing required around stairwell opening and on stairs

floor framing

soil around exterior basement stairwell should slope away to reduce water pressure on side walls

basement door

check this wall for bowing or cracks

basement floor drain

to sewer

4" to 6" threshold required

perspective

old footing location

drain must be present

footing should be lowered below and to either side of stairwell

cross section

0486

Insulated exterior basement stairwell

inside | outside

railing required around stairwell opening and on stairs

floor framing

soil around exterior basement stairwell should slope away to reduce water pressure on side walls

basement door

footing

check this wall for bowing or cracks

basement floor drain

drain

to sewer

4" to 6" threshold required

perspective

layer of rigid insulation to prevent frost heaving of house footing (footing doesn't need to be lowered)

insulation should also extend up sides of stairwell and extend out horizontally for about 4' around stairwell

cross section

0487

Perimeter drainage tile

existing drainage tile

the perimeter drainage tile should be extended out around the basement walkout when it is built

unfortunately, this is often not done and the tile is simply cut - leading to significant water accumulations next to the walkout and potential frost damage

existing drainage tile

0488

No outside traps

inside | outside

floor framing

basement door

footing

drain

to sewer

outside trap can freeze in winter

basement floor drain

drain

connect walkout drain to basement floor drain trap

cross section

0489

Door threshold

4" to 6" threshold

drain

drains can get clogged with snow, ice or leaves (or may not be able to keep up with a heavy rainfall) - a 4" to 6" threshold is needed to prevent water from entering under the door

0490

Walkout cover or roof

walkout with roof

walkout with cover

0491

Recommended grading slopes

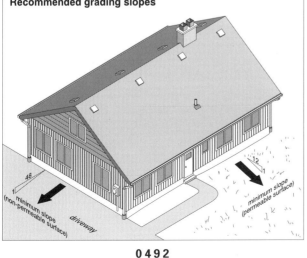

48 1 minimum slope (non-permeable surface)

driveway

12 1 minimum slope (permeable surface)

0492

Water penetration due to subgrade soil conditions

even though the grading around the house appears appropriate, the downspout discharge percolates down to the clay layer and collects next to the foundation where it may find its way into the building

wall framing

downspout

floor framing

foundation wall

porous soil (e.g. sand)

water flow

impervious soil (e.g. clay)

basement

perimeter drainage tile

0493

Swales

when the overall lot drainage is toward the house, swales can be used to direct surface water away from the foundation

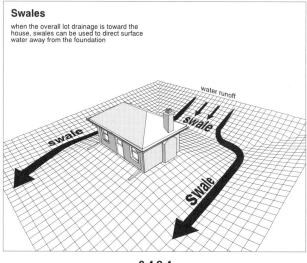

water runoff

swale

swale

swale

0494

Maximum slopes

watch for cracks in side walls due to soil movement

bent tree trunks can indicate soil slippage

vegetation helps reduce erosion from rain runoff and improves the stability of the top soil layer

water at the bottom of the slope can erode the soil, leading to increased slope instability

2
1

the maximum slope for a lot should be 1 in 2

water

0495

Evidence of exterior insulation installation

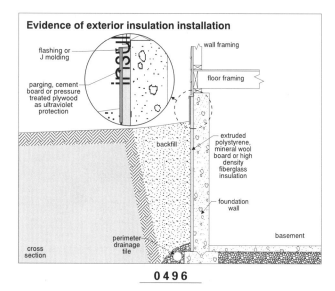

flashing or J molding

wall framing

floor framing

parging, cement board or pressure treated plywood as ultraviolet protection

extruded polystyrene, mineral wool board or high density fiberglass insulation

backfill

foundation wall

cross section

perimeter drainage tile

basement

0496

Drainage layer

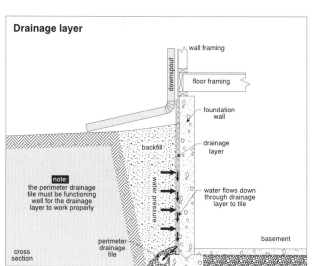

wall framing

downspout

floor framing

foundation wall

backfill

drainage layer

note:
the perimeter drainage tile must be functioning well for the drainage layer to work properly

water pressure

water flows down through drainage layer to tile

basement

cross section

perimeter drainage tile

0497

Foundation cracks - repair methods

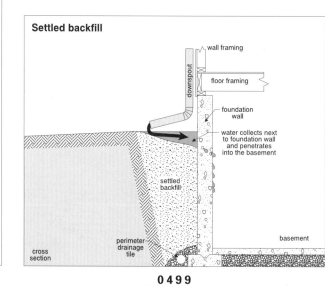

wall framing

floor framing

foundation wall

wall framing

floor framing

wooden "box"

finish grade

drainage material

(A) wooden "box" holds soil away from crack and permits free drainage down to drainage tile

(B) layer of drainage material to keep water pressure away from crack

0498

Settled backfill

wall framing

downspout

floor framing

foundation wall

water collects next to foundation wall and penetrates into the basement

settled backfill

cross section

perimeter drainage tile

basement

0499

French drain

downspout

below-grade
drainage pipe
from downspout

french
drain

french drain should be
at least 15 feet away
from the house

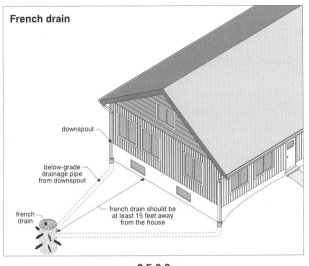

0 5 0 0

Excavation, dampproofing and drainage tile

cross section showing
water draining through
drainage material

drainage
material

soil

foundation

water

drainage
material

drainage
tile

dampproofing

gravel

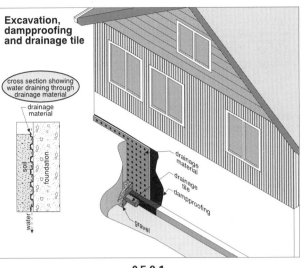

0 5 0 1

Water penetration due to subgrade soil conditions

even though the grading around the house
appears appropriate, the downspout discharge
percolates down to the clay layer and collects
next to the foundation where it may find its way
into the building

wall framing

downspout

floor framing

foundation
wall

porous
soil
(e.g. sand)

water flow

impervious soil (e.g. clay)

basement

perimeter
drainage
tile

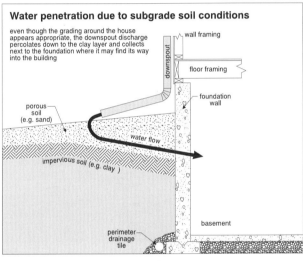

0 5 0 2

Interior flat roof drains

when flat roofs are
surrounded by parapet
walls, interior roof drains
are commonly used

parapet
wall

drainage

drain

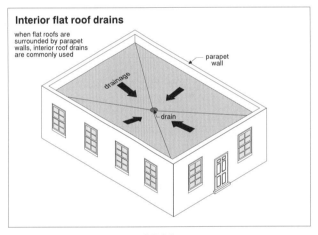

0 5 0 3

Integral gutter

roof
sheathing

shingles

roof
rafter

attic space

gutter lining
(typically sheet metal)

gutter

ceiling joist

fascia

wall
sheathing

wall
stud

gutter leakage can
damage structural
components

cross
section

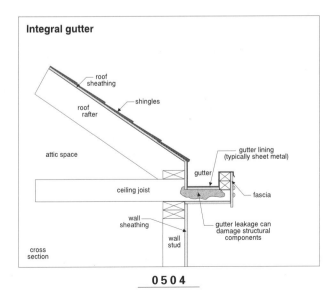

0 5 0 4

Yankee gutter

top view

shingles

this area prone ice
damming, moss growth
and shingle wear

drainage

shingles

roof
rafter

yankee
gutter

ceiling joist

wall
stud

wall
sheathing

soffit

shingles

yankee
gutter

roof
sheathing

cross
section

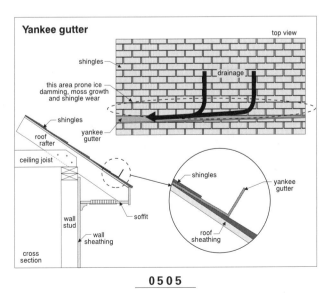

0 5 0 5

Identifying aluminum and galvanized steel gutters

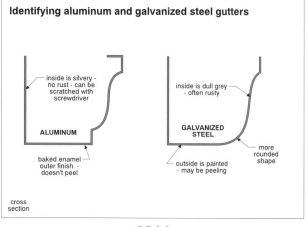

inside is silvery - no rust - can be scratched with screwdriver

ALUMINUM

baked enamel outer finish - doesn't peel

inside is dull grey - often rusty

GALVANIZED STEEL

outside is painted - may be peeling

more rounded shape

cross section

0506

Excess shingle overhang

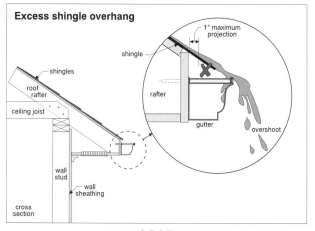

shingles

roof rafter

ceiling joist

wall stud

wall sheathing

1" maximum projection

shingle

rafter

gutter

overshoot

cross section

0507

Gutters - common reasons for leakage

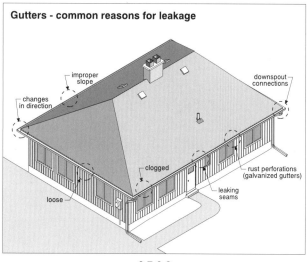

improper slope

changes in direction

downspout connections

clogged

loose

leaking seams

rust perforations (galvanized gutters)

0508

Repairing an integral gutter

if the original gutter lining has deteriorated (rust or leaking seams), a new liner made of single (or double) ply roofing material can be applied over top

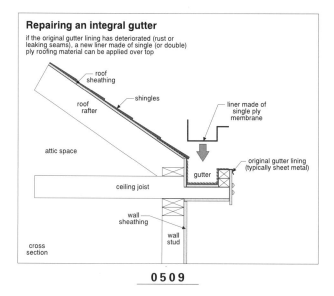

roof sheathing

roof rafter

shingles

attic space

liner made of single ply membrane

original gutter lining (typically sheet metal)

gutter

ceiling joist

wall sheathing

wall stud

cross section

0509

Gutter attachment methods

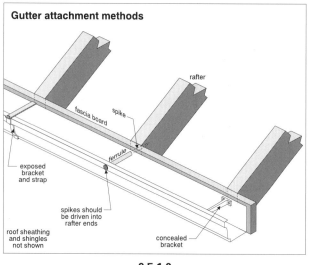

rafter

fascia board

spike

ferrule

exposed bracket and strap

spikes should be driven into rafter ends

concealed bracket

roof sheathing and shingles not shown

0510

Screens on gutters

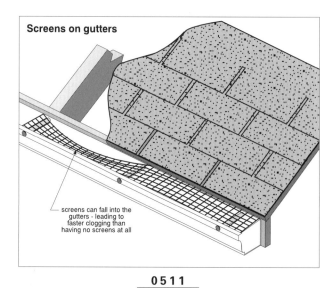

screens can fall into the gutters - leading to faster clogging than having no screens at all

0511

Gutter and downspout installation

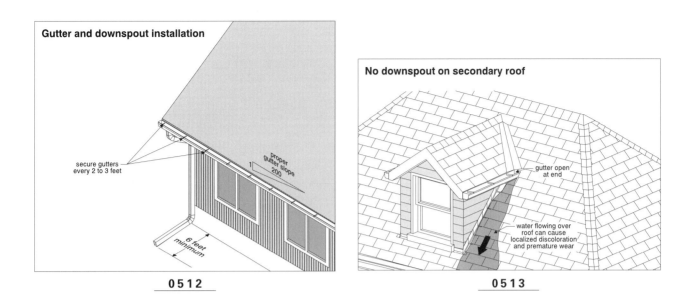

secure gutters every 2 to 3 feet

proper gutter slope 1 / 200

6 feet minimum

0 5 1 2

No downspout on secondary roof

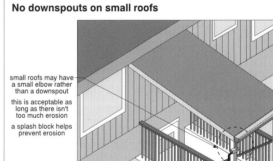

gutter open at end

water flowing over roof can cause localized discoloration and premature wear

0 5 1 3

Downspout running across roof

installing a downspout (from the secondary roof to the main gutter below) helps prevent localized roof wear

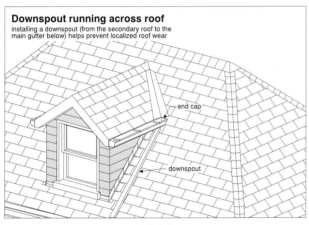

end cap

downspout

0 5 1 4

No downspouts on small roofs

small roofs may have a small elbow rather than a downspout

this is acceptable as long as there isn't too much erosion

a splash block helps prevent erosion

concrete splash block

0 5 1 5

Downspouts - common leakage areas

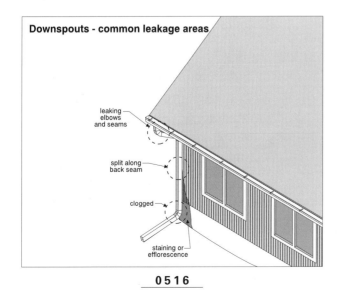

leaking elbows and seams

split along back seam

clogged

staining or efflorescence

0 5 1 6

Downspouts - proper connection of sections

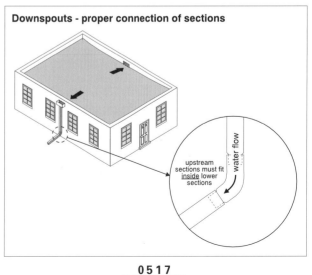

water flow

upstream sections must fit <u>inside</u> lower sections

0 5 1 7

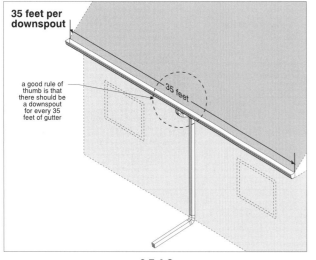

35 feet per downspout

a good rule of thumb is that there should be a downspout for every 35 feet of gutter

35 feet

0518

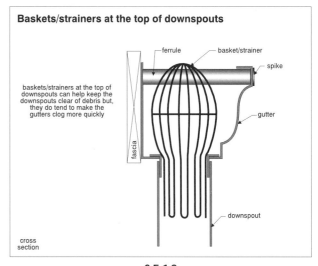

Baskets/strainers at the top of downspouts

ferrule

basket/strainer

spike

fascia

gutter

baskets/strainers at the top of downspouts can help keep the downspouts clear of debris but, they do tend to make the gutters clog more quickly

downspout

cross section

0519

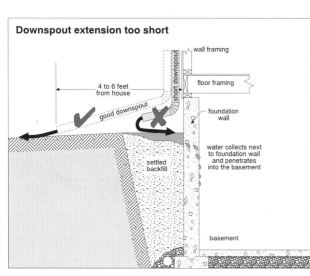

Downspout extension too short

wall framing

short downspout

4 to 6 feet from house

good downspout

floor framing

foundation wall

water collects next to foundation wall and penetrates into the basement

settled backfill

basement

0520

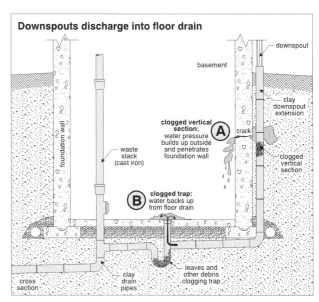

Downspouts discharge into floor drain

downspout

basement

clay downspout extension

clogged vertical section:
water pressure builds up outside and penetrates foundation wall

Ⓐ crack

clogged vertical section

foundation wall

waste stack (cast iron)

Ⓑ clogged trap:
water backs up from floor drain

leaves and other debris clogging trap

clay drain pipes

cross section

0521

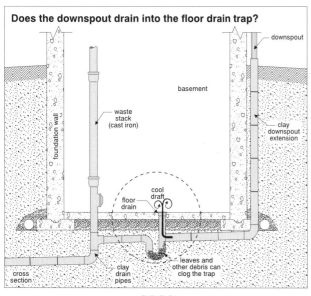

Does the downspout drain into the floor drain trap?

downspout

basement

clay downspout extension

foundation wall

waste stack (cast iron)

cool draft

floor drain

leaves and other debris can clog the trap

clay drain pipes

cross section

0522

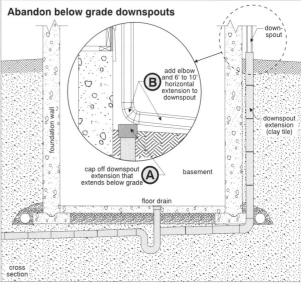

Abandon below grade downspouts

down-spout

add elbow and 6' to 10' horizontal extension to downspout

Ⓑ

foundation wall

cap off downspout extension that extends below grade

Ⓐ

downspout extension (clay tile)

basement

floor drain

cross section

0523

Scupper drains

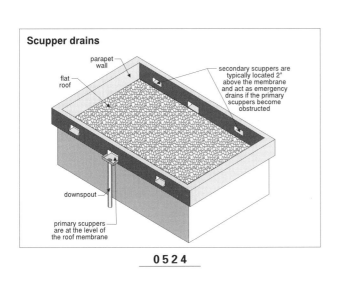

parapet wall

flat roof

secondary scuppers are typically located 2" above the membrane and act as emergency drains if the primary scuppers become obstructed

downspout

primary scuppers are at the level of the roof membrane

0524

Weight of water

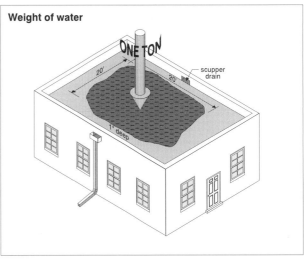

ONE TON

20'

20

scupper drain

1' deep

0525

Clogged interior flat roof drain

interior flat roof drains clogged with debris can cause water ponding, leading to: shortened roof life, possible freeze/thaw damage, additional weight/roof sag and significant water infiltration if a leak occurs

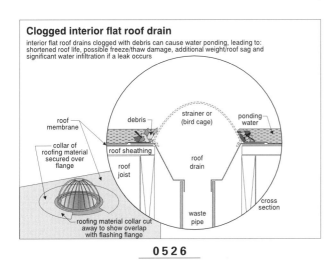

roof membrane

collar of roofing material secured over flange

debris

strainer or (bird cage)

roof sheathing

roof joist

ponding water

roof drain

waste pipe

cross section

roofing material collar cut away to show overlap with flashing flange

0526

Drains in columns and walls

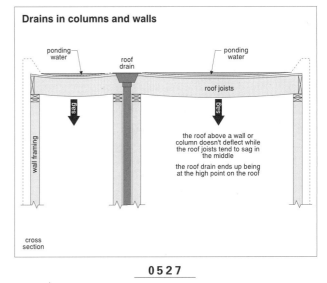

ponding water

roof drain

ponding water

roof joists

wall framing

sag

sag

the roof above a wall or column doesn't deflect while the roof joists tend to sag in the middle

the roof drain ends up being at the high point on the roof

cross section

0527

Adding a roof drain

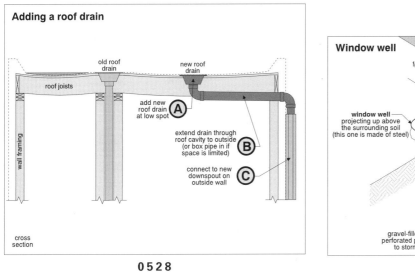

old roof drain

new roof drain

roof joists

add new roof drain at low spot **A**

extend drain through roof cavity to outside (or box pipe in if space is limited) **B**

connect to new downspout on outside wall **C**

wall framing

cross section

0528

Window well

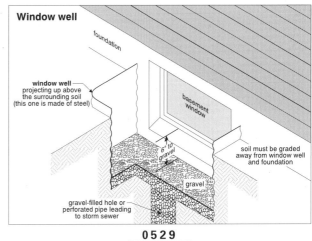

foundation

window well projecting up above the surrounding soil (this one is made of steel)

basement window

6" to gravel

soil must be graded away from window well and foundation

gravel

gravel-filled hole or perforated pipe leading to storm sewer

0529

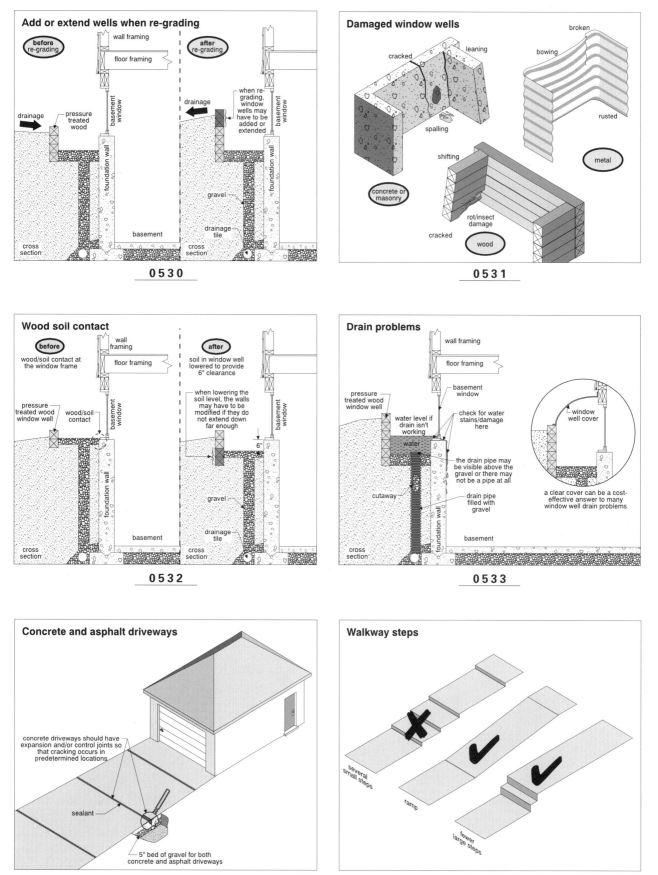

Add or extend wells when re-grading

before re-grading

after re-grading

wall framing

floor framing

drainage

pressure treated wood

basement window

foundation wall

drainage

when re-grading, window wells may have to be added or extended

basement window

foundation wall

gravel

drainage tile

basement

cross section

cross section

0530

Damaged window wells

cracked

leaning

broken

bowing

rusted

spalling

shifting

metal

rot/insect damage

cracked

concrete or masonry

wood

0531

Wood soil contact

before

after

wall framing

floor framing

wood/soil contact at the window frame

soil in window well lowered to provide 6" clearance

pressure treated wood window well

wood/soil contact

when lowering the soil level, the walls may have to be modified if they do not extend down far enough

basement window

basement window

foundation wall

6"

gravel

drainage tile

basement

cross section

cross section

0532

Drain problems

wall framing

floor framing

basement window

pressure treated wood window well

check for water stains/damage here

water level if drain isn't working

water

the drain pipe may be visible above the gravel or there may not be a pipe at all

window well cover

cutaway

drain pipe filled with gravel

a clear cover can be a cost-effective answer to many window well drain problems

foundation wall

basement

cross section

0533

Concrete and asphalt driveways

concrete driveways should have expansion and/or control joints so that cracking occurs in predetermined locations

sealant

5" bed of gravel for both concrete and asphalt driveways

0534

Walkway steps

several small steps

ramp

fewer large steps

0535

Trees and shrubs too close to house

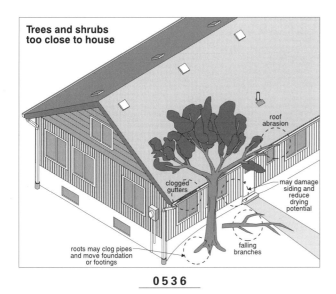

roof abrasion

clogged gutters

may damage siding and reduce drying potential

roots may clog pipes and move foundation or footings

falling branches

0 5 3 6

Raised patios

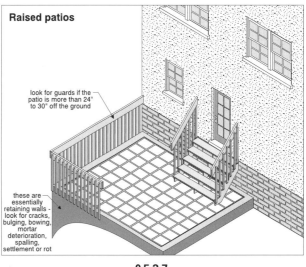

look for guards if the patio is more than 24" to 30" off the ground

these are essentially retaining walls - look for cracks, bulging, bowing, mortar deterioration, spalling, settlement or rot

0 5 3 7

Gabion retaining wall

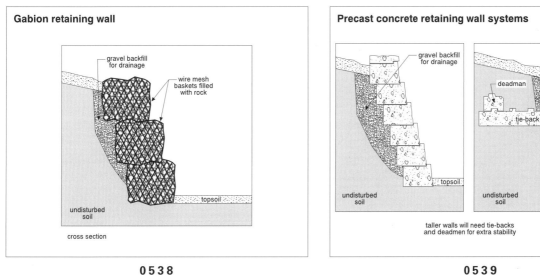

gravel backfill for drainage

wire mesh baskets filled with rock

topsoil

undisturbed soil

cross section

0 5 3 8

Precast concrete retaining wall systems

gravel backfill for drainage

gravel backfill for drainage

deadman

tie-back

topsoil

topsoil

undisturbed soil

undisturbed soil

cross section

taller walls will need tie-backs and deadmen for extra stability

0 5 3 9

Cantilevered concrete retaining wall

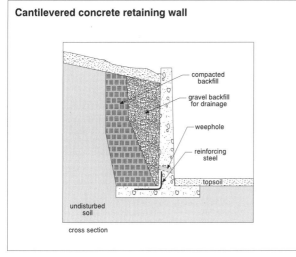

compacted backfill

gravel backfill for drainage

weephole

reinforcing steel

topsoil

undisturbed soil

cross section

0 5 4 0

Pile retaining walls (shoring)

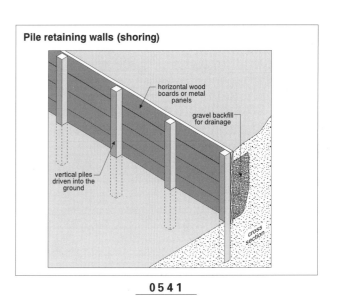

horizontal wood boards or metal panels

gravel backfill for drainage

vertical piles driven into the ground

cross section

0 5 4 1

Wood retaining wall

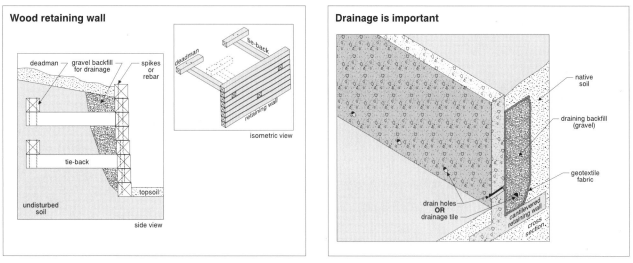

deadman
gravel backfill for drainage
spikes or rebar
tie-back
topsoil
undisturbed soil

side view

tie-back
deadman
retaining wall

isometric view

0542

Drainage is important

native soil

draining backfill (gravel)

geotextile fabric

drain holes **OR** drainage tile

cantilevered retaining wall

cross section

0543

Movement or cracking

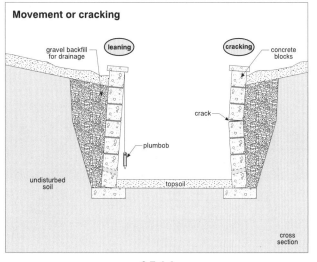

gravel backfill for drainage

leaning

cracking

concrete blocks

crack

plumbob

undisturbed soil

topsoil

cross section

0544

Block walls need rebar and cap

coping

gravel backfill for drainage

concrete blocks

weephole

reinforcing steel (rebar)

footing

topsoil

undisturbed soil

cross section

0545

Inspecting retaining walls - things to watch for

slumping

wall not plumb - leaning away

retaining wall

slumping

bottom of the wall has slipped

retaining wall

wall bowing

cracks

displace-ment of individual units

crack

displacement

0546

Weep holes in retaining wall

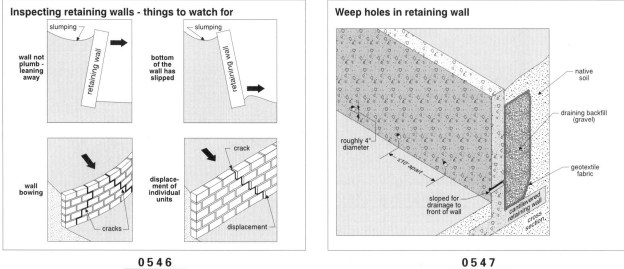

native soil

draining backfill (gravel)

geotextile fabric

roughly 4" diameter

5'0 apart

sloped for drainage to front of wall

cantilevered retaining wall

cross section

0547

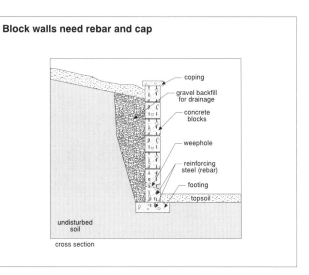

With almost 500 illustrations, *Heating and Air Conditioning* shows you exactly what the finished job should look like. This book is an excellent tool for the new home buyer or inspector, showing you which pitfalls to avoid and what to look for when searching for potential problems in the heating and air conditioning systems of your or your client's home.

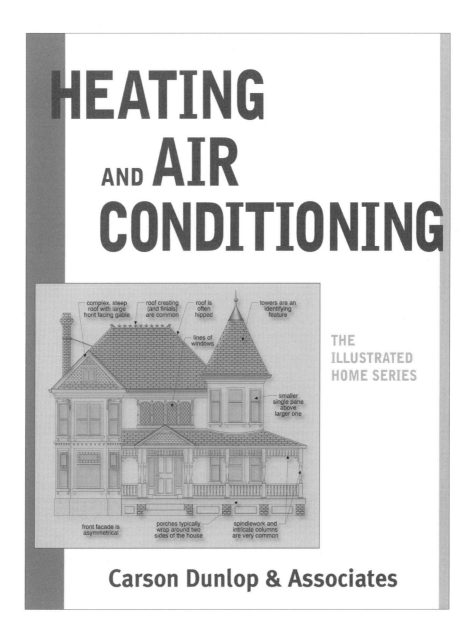

With more than 450 illustrations, *Electrical, Plumbing, Insulation and the Interior* shows you exactly what the finished job should look like. This book is an excellent tool for the new home buyer or inspector, showing you which pitfalls to avoid and what to look for when searching for potential problems with the electrical work, plumbing and insulation of your or your client's home.

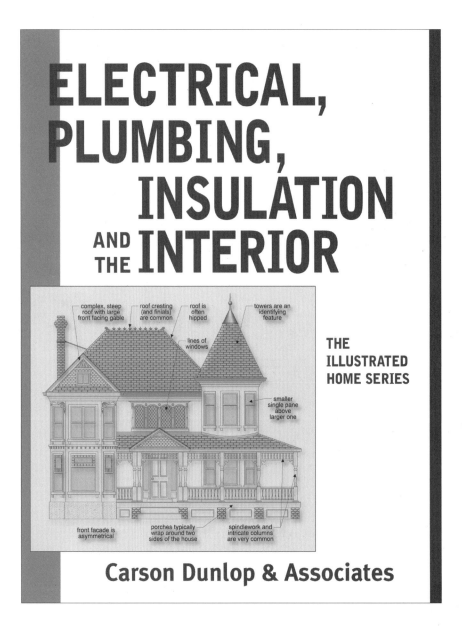

ELECTRICAL, PLUMBING, INSULATION AND THE INTERIOR

THE ILLUSTRATED HOME SERIES

complex, steep roof with large front facing gable

roof cresting (and finials) are common

roof is often hipped

towers are an identifying feature

lines of windows

smaller single pane above larger one

front facade is asymmetrical

porches typically wrap around two sides of the house

spindlework and intricate columns are very common

Carson Dunlop & Associates